The Dawn of *Dynamic* Cosmos Dimension Arrays

Dirac Equations Based on Dimension Arrays
Dirac matrices, Strong SU(4) Matrices, ElectroWeak SU(2)⊗U(1)) Matrices
Two-Tier Theory Coordinates Support for Superluminal Motion
Two-Tier Theory for Renormalization
Superluminal Model
Superluminal Starships into the Galaxy

Stephen Blaha Ph. D.
Blaha Research

Pingree-Hill Publishing
MMXXIV

.

ISBN: 979-8-9894084-8-1

Rev. 00/00/01 October 20, 2024

The Dawn of *Dynamic* Cosmos Dimension Arrays

Dirac Equations Based on Dimension Arrays
Dirac matrices, Strong SU(4) Matrices, ElectroWeak SU(2)⊗U(1)) Matrices
Two-Tier Theory Coordinates Support for Superluminal Motion
Two-Tier Theory for Renormalization
Superluminal Model
Superluminal Starships into the Galaxy

Stephen Blaha Ph. D.
Blaha Research

Pingree-Hill Publishing
MMXXIV

.

ISBN: 979-8-9894084-8-1

Rev. 00/00/01 October 20, 2024

To Margaret

Some Other Books by Stephen Blaha

SuperCivilizations: Civilizations as Superorganisms (McMann-Fisher Publishing, Auburn, NH, 2010)

All the Universe! Faster Than Light Tachyon Quark Starships & Particle Accelerators with the LHC as a Prototype Starship Drive Scientific Edition (Pingree-Hill Publishing, Auburn, NH, 2011).

Unification of God Theory and Unified SuperStandard Model THIRD EDITION (Pingree Hill Publishing, Auburn, NH, 2018).

The Exact QED Calculation of the Fine Structure Constant Implies ALL 4D Universes have the Same Physics/Life Prospects (Pingree Hill Publishing, Auburn, NH, 2019).

Passing Through Nature to Eternity ProtoCosmos, HyperCosmos, Unified SuperStandard Theory (Pingree Hill Publishing, Auburn, NH, 2022).

HyperCosmos Fractionation and Fundamental Reference Frame Based Unification: Particle Inner Space Basis of Parton and Dual Resonance Models (Pingree Hill Publishing, Auburn, NH, 2022).

The Cosmic Panorama: ProtoCosmos, HyperCosmos, Unified SuperStandard Theory (UST) Derivation (Pingree Hill Publishing, Auburn, NH, 2022).

God and and Cosmos Theory (Pingree Hill Publishing, Auburn, NH, 2023).

Newton's Apple is Now The Fermion (Pingree Hill Publishing, Auburn, NH, 2023).

Cosmos Theory: The Sub-Particle Gambol Model (Pingree Hill Publishing, Auburn, NH, 2023).

Cosmos-Universe-Particle-Gambol Theory (Pingree Hill Publishing, Auburn, NH, 2024).

Fractal Cosmos Curve: Tensor-based Cosmos Theory (Pingree Hill Publishing, Auburn, NH, 2024).

*The Eternal Form of Cosmos Theory Third Edition (*Pingree Hill Publishing, Auburn, NH, 2024).

*Fundamental Constants of Cosmos Theory and The Standard Model (*Pingree Hill Publishing, Auburn, NH, 2024).

*Geometric Cosmos Geometric Universe (*Pingree Hill Publishing, Auburn, NH, 2024).

Available on Amazon.com, bn.com Amazon.co.uk and other international web sites as well as at better bookstores.

CONTENTS

FIGURES and TABLES

Introduction

This book shows all the matrices that are present in the Dirac equations for derivative terms, SU(4) and SU(3)⊗U(1) Strong Interactions, and the SU(2)⊗U(1) Electroweak Interactions have a common origin in Cosmos Theory dimension arrays. These interactions are present throughout our universe's Unified SuperStandard Theory and also in universes of higher space-time dimension. (Cosmos universes have dimensions ranging from zero through 18.)

The natural form of the Dirac equations lend themselves to being defined with the author's Quadplex fermion wave functions presented in earlier books. The *dynamics* formalism that results appears in chapter 4. Combined with earlier books on Cosmos Theory we have complete elementary particle formalism.

We have suggested in the past that Cosmos Theory should be based on the author's Two-Tier Theory of 2002. This theory enables Perturbation Theory calculations to avoid ultra-violet infinities in our universe.

A major reason for using Two-Tier Theory is its ability to eliminate similar infinities that would appear in higher dimension universes, which are likely to be present in the Cosmos. (Other approaches to renormalization do not apply in higher dimension space-times.)

This book provides a thorough presentation of Two-Tier Theory in chapters 5 and 6.

A possible benefit of Two-Tier Theory is its use of Two-Tier coordinates which are complex-valued (although approximated with real-valued coordinates at low energies). These coordinates have an inherent oscillatory nature. They can be used in a model, presented in this book, that supports transitions between subluminal and superluminal (faster than light) motion.

A potential application of models of this type would be the development of starships for *speedy* transport to the stars at velocities well in excess of light speed. To this end the book considers: superluminal motion, potential problems related to superluminal starships, and instantaneous interstellar communication via quantum entanglement. *The Superluminal motion in this model is directly based on Two-Tier Theory.*

This book, and the author's previous books, shows the Cosmos is a compact, highly structured set of spaces, universes, and particle dynamics.

1. Structure of Cosmos Theory

This chapter outlines some of the features of Cosmos Theory. The Fundamental Reference Frame and the 42 and 88 dimension unification spaces are not discussed here. The reader is referred to our recent books of these topics.

1.1 Cosmos Theory Spaces

There are 10 physical HyperCosmos spaces. They are listed in Fig. 1.1. These spaces are numbered by even dimensions denoted r = 0, 2, 4, …, 18. They each define a square dimension array with 2^{r+4} components. In each space the dimension array has r space-time dimensions and $2^{r+4} - r$ dimensions for internal symmetries. The internal symmetries found throughout the spaces are: SU(4) (or broken to SU(3)⊗U(1)), SU(2)⊗U(1), U(1), and a space-time symmetry.

The number of physical HyperCosmos spaces, ten, is set by a consistency condition that balances the internal thermodynamic expansion pressure of a universe against the external constraining vacuum Casimir force. This condition eliminates the need for a dynamics before the creation of universes. Time does not exist before the existence of universes. There is nothing before the beginning except structure without substance.

There is also a set of spaces that we call Cosmos spaces of the Second Kind that are similar to HyperCosmos spaces. They are discussed in previous books by the author.

The 42 and 88 dimension unification spaces complete the list of Cosmos Theory spaces.

Dimension arrays grow in size from space to space by factors of 4. See the d_{dN} column in Fig. 1.1. Fig. 1.2 shows the UST dimension array populated by fundamental fermions. Fig. 1.3 shows the UST dimension array populated by symmetries.

We have found that other quantities in the Unified SuperStandard theory (UST), which has an r = 4 HyperCosmos space, also grow by various factors: coupling constants, and first generation fermion masses in two sequences.

Later we will see that γ-matrices, SU(4) matrices, and SU(2) ⊗U(1) matrices have a format similar to dimension arrays in the UST. Thus dimension arrays are integral to the UST and HyperCosmos universes in general.

This short summary of previous work by the author does not present many important details. Its purpose is to give an initial view prefatory to the later chapters.

THE Physical HYPERCOSMOS SPACES SPECTRUM

THE HYPERCOSMOS SPACES SPECTRUM

Blaha Space Number $N = o_s$	Cayley-Dickson Number n	Cayley Number Cn d_c	Dimension Array column length d_{cd}	Dimension Array Size d_{dN}	Space-time-Dimension r	CASe Group $su(2^{r/2},2^{r/2})$ CASe
0	10	1024	2048	2048^2	18	su(512,512)
1	9	512	1024	1024^2	16	su(256,256)
2	8	256	512	512^2	14	su(128,128)
3	7	128	256	256^2	12	su(64,64)
4	6	64	128	128^2	10	su(32,32)
5	5	32	64	64^2	8	su(16,16)
6	4	16	32	32^2	6	su(8,8)
7	**3**	**8**	**16**	$\mathbf{16^2}$	**4**	**su(4,4)**
8	2	4	8	8^2	2	su(2,2)
9	1	2	4	4^2	0	su(1,1)

Figure 1.1. The part of the Cosmos spectrum of spaces that can be used to define universes based on our analysis of the Cosmos consistency condition. This spectrum appears in Blaha (2022g) and our earlier books.

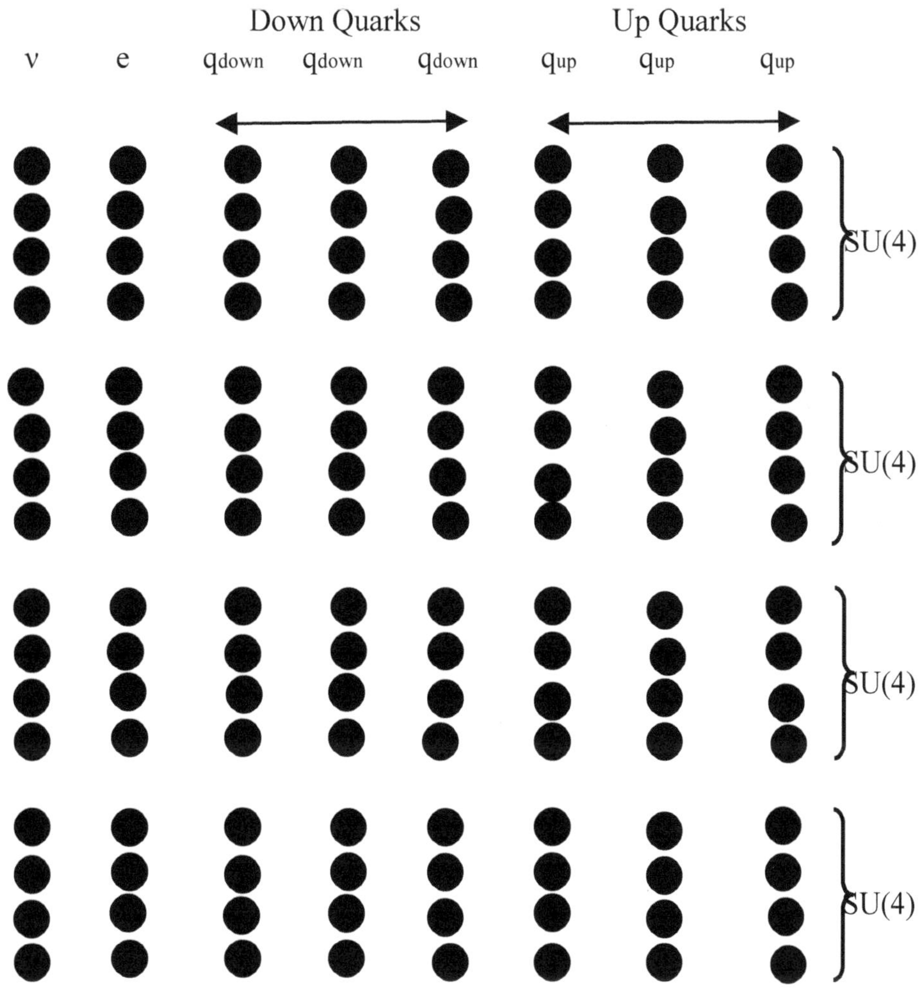

Figure 1.2. Four SU(3) Strong interaction groups in the Normal sector and four SU(3) groups in the Dark sector of the UST. Interactions are between any quark of any generation within each layer in the Normal sector and also in the Dark sector.. There is a different SU(3) for each layer.in the Normal and Dark sectors totally to 8 SU(3)'s. This diagram appears in Blaha (2023d) and our earlier books.

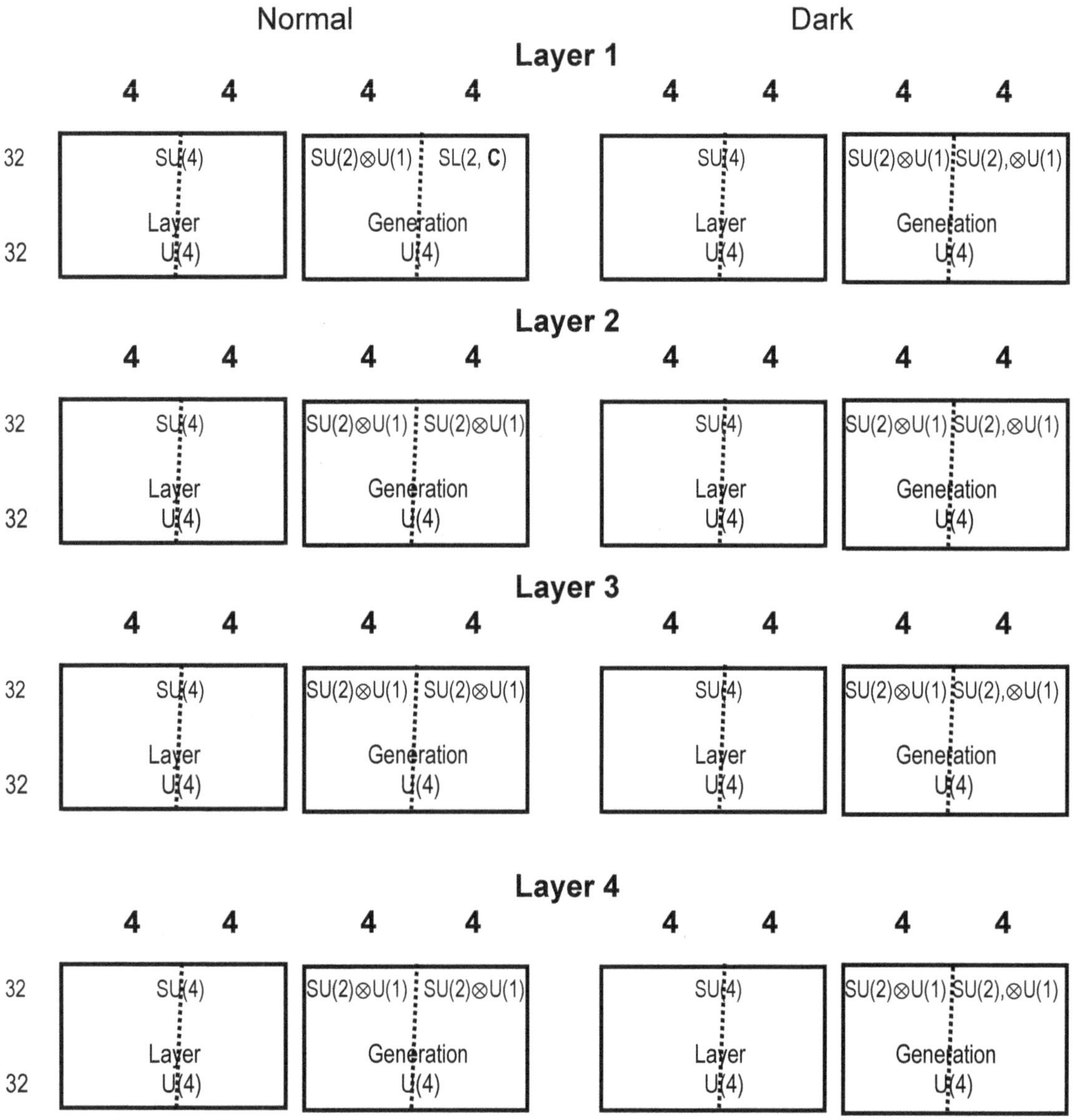

Figure 1.3. Normal and Dark symmetry groups of UST. SL(2, C) represents the Lorentz group $SO^{+}(1,3)$. This diagram appears in Blaha (2024i) and our earlier books such as Blaha (2020d). .

2. Basis of Cosmos Theory

This chapter describes the basis of Cosmos Theory in totally antisymmetric tensors and then proceeds to describe the dimension arrays that are generated.[1] The relation of Cosmos Theory arrays to fractal curves and to Dirac γ matrices are also described.

2.1 Tensor Basis

Cosmos Theory, with all its implications, is based on one mathematical fact and two assumptions. The mathematical fact follows from a consideration of the representations of the homogeneous Lorenz group in four space-time dimensions extended to other space-time dimensions r.

There is a difference between undeniable mathematical fact and *ad hoc* hypotheses. The fact in question is the number of independent totally anti-symmetric tensors of 0, 1, 2, … , r indices in r even space-time dimensions;

$$2^r \tag{2.1}$$

This number equals the number of elements in an irreducible γ-matrix.[2] The number of rows (columns) of γ-matrices is

$$2^{r/2} \tag{2.2}$$

This number is also twice the number of spins in r dimensions for the fermion of lowest spin in r dimensions.[3]

If we consider the form of a fermion's quantum fields,[4] the first assumption:

$$\begin{aligned} \psi_1(x) &= \Sigma_{\alpha,s}[b_1(\alpha, s)u_{\alpha s}f_\alpha(x) + d^\dagger{}_1(\alpha, s)v_{\alpha s}f_\alpha{}^*(x)] \\ \psi_2(x) &= \Sigma_{\alpha,s}[b_2(\alpha, s)u_{\alpha s}f_\alpha(x) + d^\dagger{}_2(\alpha, s)v_{\alpha s}f_\alpha{}^*(x)] \end{aligned} \tag{2.3}$$

where α represents the Fourier momentum, then the total number of creation and annihilation operators is

$$d_{cr} = 2^{r/2 + 2} \tag{2.4}$$

---

[1] Blaha (2023d) and (2024c).

[2] For our universe where r = 4 the γ-matrices are the Dirac γ-matrices.

[3] The least spin fermion quantum field in a space-time of dimension r has the spin $s = ½ (2^{r/2 - 1} – 1)$.

[4] In our PseudoQuantum formulation of fermions in the 1970's and recently we define two wave functions for each fermion for important reasons presented in my earlier papers and books.

where an additional factor of 2 follows from taking account of the d and $d^\dagger$ operators and a second additional factor of 2 follows due to having two quantum fields.

The column length d_{cr} specifies degrees of freedom since the associated creation and annihilation operators generate independent states.

2.2 Dimension Arrays

We may define an array, the dimension array, by introducing a group index for a group with the same number of components as d_{cr} and use it to define a dimension array with d_{dr} elements:

$$d_{dr} = d_{cr}^{\ 2} = 2^{r+4} \qquad (2.5)$$

The quantum fields with index a become:

$$\psi_{1a}(x) = \Sigma_{\alpha,s}[b_{1a}(\alpha, s)u_{\alpha s}f_\alpha(x) + d^\dagger{}_{1a}(\alpha, s)v_{\alpha s}f_\alpha{}^*(x)]$$
$$\psi_{2a}(x) = \Sigma_{\alpha,s}[b_{2a}(\alpha, s)u_{\alpha s}f_\alpha(x) + d^\dagger{}_{2a}(\alpha, s)v_{\alpha s}f_\alpha{}^*(x)]$$

The number of additional components is set equal to the number of operators, eq. 2.5, to give the group the same number of degrees of freedom. As a result d_{dr} may be represented in the form of a square array. We call it the *dimension array*. We define a space for each space-time dimension r with a corresponding dimension array. *We relate space dimensions to the set of creation/annihilation operators (with internal symmetries) since they both specify degrees of freedom.*

We make the second assumption:

We define spaces, which each have an associated dimension array that specifies a set of dimensions, and that specify the spectrum of each set of fundamental fermions, of scalar particles, and of symmetries (including space-time dimensions r and internal symmetry dimensions.)

Each space-time has an even number of dimensions. Odd space-time dimension cases are ruled out because they would have space-time dimension arrays that are redundant with the even space-time dimensions' dimension arrays. The numbers of their spins in odd dimensions are the same as the next lower even dimensions leading to redundant dimension arrays.

The dimension arrays and their associated spaces result.[5] We have called the theory of these spaces Cosmos Theory. Cosmos Theory is described in detail in our previous books.

The set of positive space-time dimension spaces ranges from r = 0 to r = ∞. We arbitrarily set the number of spaces with universes to ten. We may base this choice on a Planckian model for universes presented in Blaha (2024b).

---

[5] The appearance of powers of 2 is ultimately the result of the number of independent tensors in each space-time.

Below we show that the column lengths of dimension arrays form a sequence that mirrors the piecewise linear elements of the fractal Hilbert curve leading us to characterize the HyperCosmos spaces as forming part of the *fractal Cosmos Curve*.

2.3 Extension of Spaces to Fractional Space-time Dimensions

Cosmos spaces with fractional dimensions are part of Cosmos Theory. They area important part of the form of the Cosmos Fractal Curve.

The space-time dimensions that we have considered above are non-negative, even integer-valued dimensions greater than or equal to zero. We may introduce more dimensions by descending to negative space-time dimensions.[6] Spaces with negative integer dimensions are called the Limos spaces. They are used in gambol theory in Blaha (2024a).

We introduce positive (and now negative) Cayley-Dickson numbers n related to the positive (and negative) space-time dimensions r by

$$r = 2n - 2 \tag{2.6}$$

with the result

$$d_{cn} \equiv d_{cr} = 2^{n+1} = 2^{r/2+2} \tag{2.7}$$
$$d_{dn} \equiv d_{cr} = 2^{2n+2} = 2^{r+4} \tag{2.8}$$

We will use n to relate the dimension arrays to a Hilbert-like fractal curve.

2.4 The Fractal Cosmos Curve

We now use the fractal relations first found in Blaha (2024b). The fractal construction of the Cosmic Curve is based on the association of dimension array column lengths[7] and the orders of piecewise linear line segment lengths that are used to form a fractal curve. We relate the *Cosmos Curve* construction to the Hilbert curve construction with:[8]

Cosmos Cayley-Dickson number $n = n_H - 1$ (2.9)
Hilbert Curve Line Length $= 2^{n_H}$
Hilbert Number of "boxes" $= 2^{2n_H}$
Cosmos Dimension Array Column Length $= 2^{n+1} = 2^{n_H}$
Cosmos Dimension Array Size (number of elements in array) $= 2^{2n+2} = 2^{2n_H}$

where n is the Cosmos Cayley-Dickson number and n_H is the order in the Hilbert curve construction.

---

[6] Our ProtoCosmos model created a set of spaces with negative dimensions and fractional dimension arrays. We used them in our gambol studies (Blaha (2024a)) and to extend the Cosmos Curve to a zero dimension point. See Blaha (2024b).

[7] Order by order (dimension by dimension) the dimension array column lengths map to the length of the corresponding Hilbert curve piecewise line lengths. Blaha (2024b).

[8] The construction of the fractal curve corresponding to Cosmos Theory spaces was shown to be similar to the construction of the Hilbert fractal curve in Blaha (2024b).

The non-negative n (or r) dimension array column lengths combine to generate a two dimension filled square fractal grid from a one dimension line segment.

The negative n (r) dimension array column lengths for n = -2 through n = -∞ combine to form a one dimension line element of length 1 equal to the array column length 1 for n = -1 from a zero dimension point due to the identity

$$\sum_{n=1}^{\infty} 2^{-n} = 1 \qquad (2.10)$$

If we adjoin the dimension array column lengths for n = –∞ through n = ∞ then we have the *Cosmos Fractal Curve from a zero dimension point to a two dimension filled square grid.*

The fractal dimension generation is paralleled by the generation of dimension arrays at each step in its construction.

2.5 Euclidean Construction of Creation in Cosmos Theory

With the basis of Cosmos Theory in the tensor structure of spaces we now have a complete theory with fundamental assumptions and a construction process modeled on the fractal Cosmic Curve that parallels Euclid's formulation of Geometry. The theory is purely mathematical: a mirror of Reality.

2.6 Analogous γ-Matrix Features

The even space-time dimension γ-matrices exhibit properties similar to the Cosmos Theory dimension arrays:

1. Dimension arrays have a size that is 16 times the size of γ-matrices.

2. The γ-matrices are square. They are r matrices of size $2^{r/2}$ by $2^{r/2}$. As the even space-time dimension r increases by 2 the γ-matrices row and column sizes double. The γ-matrices quadruple as a result. For example the r = 6 γ-matrices are 8 by 8 matrices containing a quadruple of r = 4 γ-matrix parts.

3. The γ-matrices exhibit a form of nesting in quadruples.

4. Fractional γ-matrices of negative dimension n may be defined by quartering the γ-matrices of the next higher space-time dimension. These fractional matrices may be of interest for gambol quantum field theory. This topic has been considered in our previous books.

The Hilbert curve features, Cosmos dimension array features, and γ-matrix structural features are the same.

3. Dynamical Analysis of Dimension Arrays

The dynamical analysis of dimension arrays leads to a new and exciting connection between Cosmos Theory and the Quadplex unification formalism for the ElectroWeak and Strong interactions in the superluminal and subluminal quantum field theory presented in Blaha (2024j).

Below we show the need to use quadplex wave functions for fundamental particles to embody our findings within the dimension arrays for Cosmos Theory. We also need a PseudoQuantum wave function formalism in Two-Tier Theory to achieve the unification of all symmetries in all possible Cosmos universes. We defer doing that in the interests of a clear presentation since the formulation presented here brings out the ideas embodied in Cosmos dimension arrays. In addition, in the interests of clarity, we also do not use the author's Two-Tier Theory[9] in this chapter. It is required to achieve finite perturbation theory results in four dimensions (our universe) and universes of higher space-time dimension.[10] We view Two-Tier Theory[11] as the only viable way to have finite perturbation theory results to all orders in all possible universes of any dimension. Its use in Cosmos Theory makes it the only "universe friendly" *fundamental* Theory of Everything. The ramifications of this theory such as Hubble expansion of universes remain to be fully understood.

This chapter shows an intimate connection between Cosmos spaces dimension arrays, the quadplex structure of wave functions including tachyonic aspects, and the matrices (γ-matrices, SU(4) matrices, and SU(2)⊗U(1) matrices), which appear in the dynamic equations for fermions in our universe. Cosmos Theory may be said to "explain it all" for elementary particle structure and dynamics.

Cosmos dimension arrays are the template for fundamental particle matrix dynamics.

The set of Dirac- γ-matrices in our universe appear as 16 submatrices in a 256 dimension array in a quadplex formulation of fermion dynamic equations.

The 16 matrices in the SU(4) fundamental $\underline{4}$ representation appear as 16 submatrices in a 256 dimension array.

---

[99] See Blaha (2002).

[10] Our approaches to renormalization do not work in general in higher dimension quantum field theories.

[11] Chapters 5 and 6 develops aspects of Two-Tier Theory showing it effectively "adds" two imaginary dimensions to our four dimension space-time in addition to curing perturbation theory of infinities. Two-Tier Theory introduces an imaginary tachyon part in a manner that emulates duplex and quadplex bradyon-tachyon results presented here and in Blaha (2024j). While the use of Two-Tier Theory gives the conventional perturbation theory results at low energies, it gives complex results at ultra-high energies that introduce tachyonic features. These features may be the seed for superluminal (faster than light) motion that might take us to the stars.

The 4 matrices in the SU(2)⊗U(1) fundamental real-valued 4 dimension representation appear as four representations along the diagonal of a 256 dimension array.

The required quadplex form of fermion wave functions necessarily brings in bradyon-tachyon wave functions in a manner consistent with the Strong and ElectroWeak interactions. *Thus an aspect of faster than light motion – tachyons – is natural to the dynamics of Cosmos Theory matter.*

3.1 The Cosmos Dimension Arrays

Each of the ten physical Cosmos spaces (Fig. 1.1) has an associated dimension array as described in chapter 2. In a universe of each Cosmos space type the dimension array specifies the spectrum of fundamental fermions and the set of internal symmetry groups. See Figs. 1.2 and 1.3.

In this chapter we show that the dimension array of a space specifies dynamical features as well:

1. The form of the derivative term γ matrices for fermion dynamic equations (Dirac equations).

2. The forms of the SU(4) interaction terms representations. (Also SU(3)⊗U(1) after breakdown).

3. The forms of the SU(2)⊗U(1) interaction terms representations.

An important aspect of these features is that they all require the Quadplex formalism developed by this author in Blaha (2024j). This formalism combines bradyon (subluminal) and tachyon (superluminal) parts within each fundamental fermion. It opens the door to faster than light Physics. It also has an intriguing side effect: It can account for the instantaneous nature of Quantum Entanglement through the tachyon part of quadplex wave functions.

The following sections show an embedding of Quadplex features in dimension arrays in the Unified SuperStandard Theory (UST) developed by this author in recent years.

The remaining sections for the UST may be directly extended to the case of dimension arrays for higher dimension (spaces) universes such as the 6 dimension Megaverse.

3.2 Dirac-like Field Equation

For a *free* quadplex wave function in our universe we have

$$(I\ {}^{y}\gamma^{\mu}\,\partial/\partial y^{\mu} + {}^{z}\gamma^{\mu}\,\partial/\partial z^{\mu} + i\ {}^{u}\gamma^{\mu}\,\partial/\partial u^{\mu} + {}^{v}\gamma^{\mu}\,\partial/\partial v^{\mu} + M)\psi(y, z, u, v) = 0 \quad (3.1)$$

where μ is the space-time index. The equation may be reduced to subsidiary equations that are combinations of bradyon B and tachyon T subsidiary equation terms as in Blaha (2024j). The combinations are composed of parts of (B_yB_z, B_yT_z, T_yB_z, T_yT_z)(B_uB_v, B_uT_v, T_uB_v, T_uT_v) where the susbscripts indicate the y, z, u, v coordinate systems. Blaha (2024j) points out:

"We will again use the distinction between bradyons and tachyons to create a set of four Quadplex fields. Each Quadplex fermion field will consist of four parts where each part is either bradyonic or tachyonic. We associate each quadplex field with a fermion in a fundamental SU(4) representation of four dimensions.

We use b to signify a bradyonic field with b, and a tachyonic field with t. We will use the up-type and down-type[12] fermion sequences found in our previous books for quadplex fields:

Up-Type Sequence 1

$$e \leftrightarrow \psi_e = \psi_{1\uparrow} \sim b_yb_zt_ut_v \qquad (3.2)$$
$$u \leftrightarrow \psi_u = \psi_{2\uparrow} \sim b_yt_zt_ub_v$$
$$c \leftrightarrow \psi_c = \psi_{3\uparrow} \sim b_yb_zb_ut_v$$
$$t \leftrightarrow \psi_t = \psi_{4\uparrow} \sim b_yb_zb_ub_v$$

Down-Type Sequence 2

$$\nu' \leftrightarrow \psi_{\nu'} = \psi_{1\downarrow} \sim b_yt_zt_ut_v$$
$$d \leftrightarrow \psi_d = \psi_{2\downarrow} \sim b_yt_zt_ub_v$$
$$s \leftrightarrow \psi_s = \psi_{3\downarrow} \sim b_yt_zb_ut_v$$
$$b \leftrightarrow \psi_b = \psi_{4\downarrow} \sim b_yt_zb_ub_v$$

where we use particle symbols to represent wave functions with coordinates indicated with subscripts. The ElectroWeak part of each fermion uses y and z coordinates, and the SU(4) part uses u and v coordinates (together with y and z parts). All four coordinate systems have four dimensions. The top-type fermion assignment [in eq. 3.2] is *generally based on the ordering of quark masses with heavier fermions having more Bradyon parts and lighter fermions having more Tachyon parts.* A similar assignment may be made for down-type fermions. The list is modified to match ElectroWeak pairs with up-type bradyons with corresponding down-type tachyons except for the e – ν (neutrino) case where the match is e - ν'."

We will use the quadplex formalism below. *The quadplex formalism is naturally appropriate for the Cosmos Theory – UST presentation.*

3.3 Derivative Terms

Free fermion wave functions ψ(y, z, u, v) have a general form such as

$$i\, {}^y\gamma^\mu\, \partial/\partial y^\mu + {}^z\gamma^\mu\, \partial/\partial z^\mu + i\, {}^u\gamma^\mu\, \partial/\partial u^\mu + {}^v\gamma^\mu\, \partial/\partial v^\mu \qquad (3.3)$$

described in Blaha (2024j). The omission of i's follows from tachyonic-like behavior. Other forms with a different pattern of i's are also used in Blaha (2024j) as indicated in the above quote.

---

[12] Up-type particles are isospin up particles. Down-type particles are isospin down particles.

The γ matrices for the coordinate systems are shown to be embedded in the UST dimension array in Fig. 3.1. Note that the quadplex formalism is needed to "fill" the dimension array.

The γ matrices embedding in a dimension array for the UST may be directly extended to the case of dimension arrays for higher dimension (spaces) universes such as the 6 dimension Megaverse.

3.4 SU(4) Interaction Term

The Strong Interaction SU(4) group has four complex coordinates fundamental representations. We will define two representations. One representation (called sequence 1) will describe up-type fermions: ν', u, c, and t. The other representation (called sequence 2) will describe down-type fermions: e, d, s, and b. These sequences were described in Blaha (2024i). The ν' fermion is a heavy neutrino corresponding to the first generation ν. It has not yet been found in Nature. The separation of the set of fermions into two sequences is based in part on their differing electric charges with down-type fermions one charge unit below their corresponding up-type fermion.

We provisionally define the SU(4) term in the fermion field equation with

$$g_s\, A^{a\mu}(y)\, {}^y\gamma_\mu T_a \tag{3.4}$$

where the T_a are 4×4 SU(4) matrices, ${}^y\gamma_\mu$ are Dirac matrices for the y coordinate space, g_s is the SU(4) coupling constan, and $A^{a\mu}(y)$ represents SU(4) vector boson fields. (This definition may be generalized to other coordinate systems. For example, $A^{a\mu}(y, z, u, v)$ is a theoretic possibility although this type of extension is not supported by experiment.)

The SU(4) interaction term becomes different if we use the bradyon-tachyon formulation of SU(4) interactions within the framework of a bradyon-tachyon formalism developed in chapter 9 of Blaha (2024i). In this theory fermions are endowed with bradyon and tachyon parts. Eq. 3.2 shows the y, z, u and v coordinate system parts of fermions. Wave functions thus have transitions between bradyon and tachyon parts under SU(4) and ElectroWeak transformations. Thus the SU(4) matrices must have factors within them producing bradyon-tachyon changes. Therefore the SU(4) matrices are replaced by $[S_{kij}]$ matrices:

$$S_{kij} = T_{kij} T^S{}_{ij} \tag{3.5}$$

Under SU(4) u and v based bradyon-tachyon changes we found

$$T^S = T^S{}_\uparrow = T^S{}_\downarrow = \begin{bmatrix} 1 & S_v & S_u & S_uS_v \\ S_v^{-1} & 1 & S_uS_v^{-1} & S_u \\ S_u^{-1} & S_u^{-1}S_v & 1 & S_v \\ S_u^{-1}S_v^{-1} & S_u^{-1} & S_v^{-1} & 1 \end{bmatrix} \tag{3.6}$$

in Blaha (2024i) based on its section 3.4.3 where

$$S_u\psi(u) = \psi_T(u)$$
$$S_u^{-1}\psi_T(u) = \psi(u)$$
$$S_v\psi(v) = \psi_T(v)$$
$$S_v^{-1}\psi_T(v) = \psi(v)$$

maps between bradyons and tachyons. The minimal transformations are[13]

$$S_u = i\gamma_u^{\ 0}\gamma_{u3}$$
$$S_v = i\gamma_v^{\ 0}\gamma_{v3}$$
$$S_u^{-1} = -i\gamma_u^{\ 0}\gamma_{u3}$$
$$S_v^{-1} = -i\gamma_v^{\ 0}\gamma_{v3}$$

Fig. 3.2 shows the SU(4) T_k matrices embedded within a UST dimension array. *Note that the quadplex formalism is needed to "fill" the dimension array.*

The T matrices embedded in a dimension array for the UST may be directly extended to the case of dimension arrays for higher dimension (spaces) universes such as the 6 dimension Megaverse. This embedding is facilitated by the embedding of the dimension r = 4 dimension array in quadruplicate in the r = 6 dimension array.

3.5 Dimension Array Quadplex SU(4) S Matrices

The quadplex form of the SU(4) interaction using the S_a matrices is

$$g_s\, A^{a\mu}(y)\, {}^y\gamma_\mu S_a \qquad (3.7)$$

Figs. 3.4 and 3.5 show the 8 dimension form of the 8 first generation (first layer) UST fermions and their view for the Strong Interactions. Similar representations would hold for the other three generations. The other three UST layers would also have similar representations. These forms also describe the SU(3)⊗U(1) view after breakdown.

3.6 ElectroWeak SU(2)⊗U(1) Interaction Terms

The ElectroWeak Interaction SU(2)⊗U(1) group (and interaction) has a two complex coordinates fundamental representation. The fundamental representation may also be represented by a 4 real-valued dimension set of matrices. The set of these matrices appear in Fig. 3.3.

The two complex coordinates fundamental representations contain a corresponding pair of up-type and down-type fermions. They were described in Blaha (2024i). Each of these representations have a pair of fermion fields:

---

[13] Section 9.2 of Blaha (2024i).

$$\psi_{EW}(y, z) = \begin{bmatrix} \psi_{T_z}(y, z, u, v) \\ \\ \psi(y, z, u, v) \end{bmatrix}$$

with the top component having a tachyonic z wave function *part* (consistent with the ordering of Figs. 3.6 and 3.7, and the bottom component having a bradyonic z wave function *part*. The combined wave function ElectroWeak view for the top UST generation (top layer) of eight fermions seen in Fig. 3.7 can be represented by:

$$\Psi_{EW} = (\psi_{EW1}(y, z), \psi_{EW2}(y, z), \psi_{EW3}(y, z), \psi_{EW4}(y, z))$$

numbered by their order in the diagonal matrix. We define the form of ElectroWeak interaction terms part in the fermion field equation with[14]

$$+ ig_{EW}\mathbf{t}(S_z)W^{\mu}(y) + ig_{EW}'t_0(S_z)W_0^{\mu}(y) \tag{3.8}$$

using Pauli-type S_{EW} matrices where $W^{a\mu}(y)$ represents the SU(2)⊗U(1) vector boson fields, and g_{EW} and g_{EW}' are the coupling constants.

The Pauli matrices that appear in the ElectroWeak Lagrangian part must be enhanced due to the differing forms of the fermion fields: bradyon charged leptons and tachyon neutral leptons.

The enhanced $\mathbf{t}(S_z)$ and $t_0(S_z)$ Pauli matrices depend on transformations between the bradyon and tachyon z part of wave function pairs. They have the form

$$t_- = ½ \begin{bmatrix} 0 & S_z\gamma^5 \\ 0 & 0 \end{bmatrix} \tag{6.39}$$

$$= ½ \begin{bmatrix} 0 & i\gamma_z^{\,0}\gamma_{z3}\gamma_z^{\,5} \\ 0 & 0 \end{bmatrix}$$

$$t_+ = ½ \begin{bmatrix} 0 & 0 \\ S_z & 0 \end{bmatrix} \tag{6.40}$$

[14] Using the equation numbers of Blaha (2024i).

$$= \tfrac{1}{2}\begin{bmatrix} 0 & 0 \\ i\gamma_z^{\ 0}\gamma_{z3} & 0 \end{bmatrix}$$

$$t_3 = \tfrac{1}{2}\begin{bmatrix} -i\gamma_z^{\ 0}\gamma_{z3}\gamma_z^{\ 5} & 0 \\ 0 & i\gamma_z^{\ 0}\gamma_{z3} \end{bmatrix} \qquad (6.41)$$

$$\equiv \tfrac{1}{2}\begin{bmatrix} -1 & 0 \\ 0 & 1 \end{bmatrix}$$

$$t_0 = \tfrac{1}{2}\begin{bmatrix} 1 & 0 \\ 0 & 1 \end{bmatrix} \qquad (6.42)$$

using

$$S_z = i\gamma_z^{\ 0}\gamma_{z3}$$
$$S_z^{-1} = -i\gamma_z^{\ 0}\gamma_{z3}$$

The electric charge is

$$Q = t_0 + t_3 \qquad (6.43)$$

and the interaction terms for each of the four representations in Fig. 3.7 become

$$\mathbf{t}(S_z)\cdot W^i(y) \rightarrow t_- W^+ + t_+ W^- + t_3 W^3 \qquad (3.8)$$
$$t_0(S_z) W_0^{\ 0}(y) \rightarrow t_0 W^0$$

Fig. 3.7 shows the four complex two dimension SU(2)⊗U(1) representations embedded, in diagonal form. There are four pairs of fermions within the eight fermions in the first UST layer.

Each column, of the four columns, in Fig. 3.3 corresponds to a representation of SU(2)⊗U(1) matrices denoted with an upper index. The four entries in each column are its SU(2)⊗U(1) 4 × 4 submatrices denoted with lower indices for the real-valued SU(2)⊗U(1) representations.

The three other UST layers have different SU(2)⊗U(1) groups and representations. The submatrices of these groups have the same form as the Fig. 3.7 submatrices in the ElectroWeak view.

The matrices embedding in a dimension array for the UST may be directly extended to the case of dimension arrays for higher dimension (spaces) universes such as the 6 dimension Megaverse. This embedding is facilitated by the embedding of the dimension r = 4 dimension array in quadruplicate in the r = 6 dimension array.

***Note** that the quadplex formalism is needed to "fill" the dimension array with* **t**(S_z) and $t_0(S_z)$ *matrices in Fig. 3.3.*

3.7 Mass Terms

The sequences listed in eq. 3.2 were first found in an examination of physical fundamental fermion masses in the first generation fermions in the UST. The sequences are described in detail in Blaha (2024i).

The sequences and their masses are

	Sequence 1				Sequence 2			
	e	u	c	t	ν'	d	s	b
Mass (GeV/c^2):	0.511×10^{-3}	1.80×10^{-3}	1.28	171	8.4×10^{-5}	4.24×10^{-3}	102×10^{-3}	4.34
Multiplier:	$2^5\pi$	$2^5\pi$	$2^5\pi$		$2^5 = 32$	$2^5 = 32$	$2^5 = 32$	

Each sequence has a multiplicative relation between the masses: 2^5 for sequence 2 and $2^5\pi$ for sequence 1.

The Strong SU(4) view diagonal Lagrangian fermion mass term appears in Fig. 3.8. The ElectroWeak SU(2)⊗U(1) view diagonal Lagrangian fermion mass term view appears in Fig. 3.9.

3.8 General Form of the Strong Interaction Lagrangian

The form of the 8 dimension matrix composed of the pair of 4 dimension fermion representations $\mathcal{L}_{Strong}$ of the Strong Interaction SU(4) Lagrangian appears in Fig. 3.5. (The ElectroWeak Lagrangian is discussed separately in section 3.9 to avoid a confusing use of wave function and interaction indices.) The Strong Lagrangian part is:

$$\bar{\Psi}[i\, {}^{y}\gamma^{\mu}\, \partial/\partial y^{\mu} + {}^{z}\gamma^{\mu}\, \partial/\partial z^{\mu} + i\, {}^{u}\gamma^{\mu}\, \partial/\partial u^{\mu} + {}^{v}\gamma^{\mu}\, \partial/\partial v^{\mu}]\Psi + \mathcal{L}_{Mass} + \mathcal{L}_{Strong} \tag{3.9}$$

with Ψ defined in Fig. 3.4, $\mathcal{L}_{Mass}$ defined in Fig. 3.8, and $\mathcal{L}_{Strong}$ defined in eq. 3.5.

3.9 Form of the ElectroWeak Interaction Lagrangian Terms

The ElectroWeak Lagrangian view can be put in an eight dimension matrix form. We define an 8-vector representing the up-type and down-type sequences of fermions ordered to be in ElectroWeak pairs as in Fig. 3.6.

$$\bar{\Psi}_{EW}[i\, {}^{y}\gamma^{\mu}\, \partial/\partial y^{\mu} + {}^{z}\gamma^{\mu}\, \partial/\partial z^{\mu} + i\, {}^{u}\gamma^{\mu}\, \partial/\partial u^{\mu} + {}^{v}\gamma^{\mu}\, \partial/\partial v^{\mu}]\Psi_{EW} + \mathcal{L}_{MassEW} + \mathcal{L}_{ElectroWeak} \tag{3.10}$$

with Ψ_{EW} defined in Fig. 3.6, $\mathcal{L}_{MassEW}$ defined in Fig. 3.9, and $\mathcal{L}_{ElectroWeak}$ defined in Fig. 3.7.

The ElectroWeak Lagrangian is symbolized by a diagonal matrix of 2×2 submatrices in Fig. 3.7. These submatrices correspond, submatrix by submatrix, with the order of the fermion wave function vector of Fig. 3.6.

3.10 Cosmos Theory Dimension Arrays Dovetail with Quadplex Wave Functions

The preceding discussion, and the figures that follow, show that the UST dimension array for our universe is compatible with ElectroWeak SU(2)⊗U(1) and Strong Interaction SU(4) (and the broken SU(3)⊗U(1)) formalisms – Figs. 3.2 – 3.9. One sees the SU(4) fundamental representation matrices fit directly in the 256 dimension UST dimension array. One also sees the set of SU(2)⊗U(1) fundamental representation $\underline{4}$ matrices fit directly in the 256 dimension UST dimension array.

Thus we may regard the quadplex (and duplex) wave function formalisms, which support tachyon physics, as bringing tachyonic behavior to Cosmos Theory. The Lagrangian terms for the Strong and ElectroWeak Interactions may be viewed as directly based on Cosmos Theory dimension diagrams. Cosmos dimension arrays are the template for fundamental particle dynamics.

The quadplex and duplex formalisms for the UST in 4 dimensions generalize directly to higher Cosmos dimensions due to the quadrupling mechanism within HyperCosmos as dimension increases by twos. Thus tachyonic behavior exists in all universes of all Cosmos Theory spaces.

${}^{y}\gamma^{0}$	${}^{z}\gamma^{0}$	${}^{u}\gamma^{0}$	${}^{v}\gamma^{\mu 0}$
${}^{y}\gamma^{1}$	${}^{z}\gamma^{1}$	${}^{u}\gamma^{1}$	${}^{v}\gamma^{1}$
${}^{y}\gamma^{2}$	${}^{z}\gamma^{2}$	${}^{u}\gamma^{2}$	${}^{v}\gamma^{2}$
${}^{y}\gamma^{3}$	${}^{z}\gamma^{3}$	${}^{u}\gamma^{3}$	${}^{v}\gamma^{3}$

Figure 3.1. The dimension array for the γ matrices of the derivative terms. There is one subarray for each space-time index of the r = 4 UST of our universe. The y, z, u, and v indices indicate the four quadplex coordinate systems. Each of the 16 γ submatrices is a 4 × 4 Dirac γ matrix. The ${}^{y}\gamma^{\mu}$ matrix is for our universe with μ = 0, 1, 2, and 3.

T^1	T^2	T^3	T^4
T^5	T^6	T^7	T^8
T^9	T^{10}	T^{11}	T^{12}
T^{13}	T^{14}	T^{15}	I

Figure 3.2. The dimension array for the 16 U(4) T_k matrices of the SU(4) interaction terms of the r = 4 UST of our universe. Each of 15 submatrices is a 4 × 4 SU(4) matrix. The 16th submatrix is the identity matrix. These matrices are for the first UST layer. Other UST layers have different SU(4) groups. The submatrices of these groups have the same form as the above submatrices.

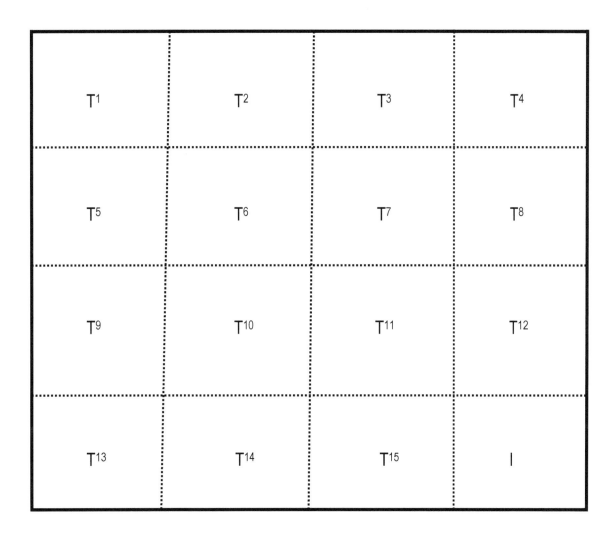

$T_{EW1}{}^{1}$	$T_{EW1}{}^{2}$	$T_{EW1}{}^{3}$	$T_{EW1}{}^{4}$
$T_{EW2}{}^{1}$	$T_{EW2}{}^{2}$	$T_{EW2}{}^{3}$	$T_{EW2}{}^{4}$
$T_{EW3}{}^{1}$	$T_{EW3}{}^{2}$	$T_{EW3}{}^{3}$	$T_{EW3}{}^{4}$
$T_{EW4}{}^{1}$	$T_{EW4}{}^{2}$	$T_{EW4}{}^{3}$	$T_{EW4}{}^{4}$

Figure 3.3. The four entries in each column are SU(2)⊗U(1) 4 × 4 submatrices denoted with lower indices in a 4 real-valued dimension SU(2)⊗U(1) representation. They are for the first UST layer. The other UST layers have different SU(2)⊗U(1) groups and representations. The submatrices of these groups have the same form as the above submatrices.

$$\Psi = \begin{bmatrix} e \\ u \\ c \\ t \\ \nu' \\ d \\ s \\ b \end{bmatrix}$$

Figure 3.4. The wave functions of the eight first layer UST fundamental fermions for use in the Strong Lagrangian view ordered by their sequences and positions within the sequences. The wave functions are denoted by the fermion's acronym.

$$\mathscr{L}_{\text{Strong}} = \bar{\Psi} \left[\begin{array}{c:c} \begin{matrix}\text{SU(4)} \\ \text{Terms}\end{matrix} & 0 \\ \hdashline 0 & \begin{matrix}\text{SU(4)} \\ \text{Terms}\end{matrix} \end{array}\right] \Psi$$

Figure 3.5. Eight dimension representation view of the pair of four dimension SU(4) representations of the fermion sequences.

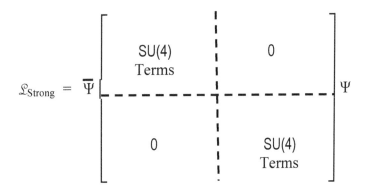

$$\Psi_{EW} = \begin{bmatrix} \nu' \\ e \\ d \\ u \\ s \\ c \\ b \\ t \end{bmatrix}$$

Figure 3.6. Fermion wave functions terms ordered as four sets of fermion wave function pairs for ElectroWeak view in Fig. 3.7.

$$\mathcal{L}_{\text{ElectroWeak}} = \overline{\Psi}_{EW} \left[\begin{array}{c|c|c|c} SU(2)\otimes U(1) & & & \\ \hline & SU(2)\otimes U(1) & & \\ \hline & & SU(2)\otimes U(1) & \\ \hline & & & SU(2)\otimes U(1) \end{array}\right] \Psi_{EW}$$

Figure 3.7. ElectroWeak Lagrangian view symbolized by a diagonal matrix of 2 × 2 submatrices. These submatrices correspond, submatrix by submatrix, with the order of the fermion wave function vector of Fig. 3.6.

$$\mathcal{L}_{\text{Mass}} = \overline{\Psi} \left[\begin{array}{cccccccc} e & & & & & & & \\ & u & & & & & & \\ & & c & & & & & \\ & & & t & & & & \\ \hline & & & & \nu' & & & \\ & & & & & d & & \\ & & & & & & s & \\ & & & & & & & b \end{array}\right] \Psi$$

Figure 3.8. Diagonal Lagrangian fermion mass matrix for Strong SU(4) view of masses. Masses denoted by their fermion symbols.

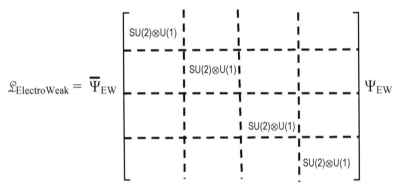

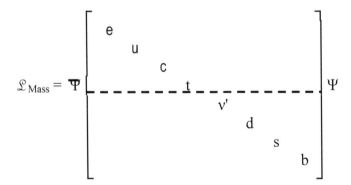

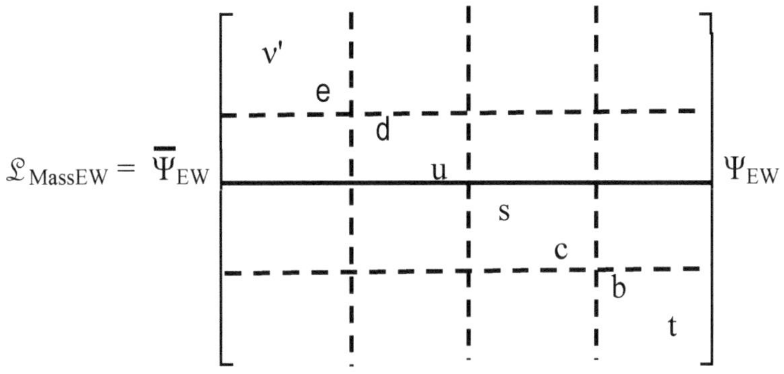

Figure 3.9. Diagonal Lagrangian fermion mass matrix for the ElectroWeak Lagrangian view of masses denoted by their fermion symbols. Other components are zeroes (not shown).

4. Two-Tier Theory, Superluminal Velocity and Cosmos Theory

This chapter considers the implications of Cosmos Theory when its universe studies use Two-Tier[15] Quantum Theory in perturbation theory quantum calculations. We will see that the Two-Tier formulation gives conventional perturbation theory results at lower energies as shown in chapter 5.[16] At ultra-high high energies, in perturbation theory studies of the interactions of matter the results display new features based on the impact of two "imaginary" dimensions "added" to our four dimension space-time. In addition to curing perturbation theory of infinities,[17] Two-Tier Theory effectively introduces an imaginary two dimension tachyon part in a manner that emulates duplex and quadplex bradyon-tachyon results presented here and in Blaha (2024j).

This ultra-high energy feature may be the seed for superluminal (faster than light) motion that might take us to the stars. An amalgam of these interactions may generate a complex propulsive force under the right conditions. This force will cause motion in our real space-time dimensions AND motion in the other two imaginary dimensions. The net velocities that could be generated in this manner will be complex and thus avoid the singularity at v = c that would otherwise prevent a transition to faster than light velocities.

4.1 Basic Two-Tier Theory Features

Our earlier books considered a coordinate system labeled X^μ that had a real-valued c-number part and an imaginary-valued q-number part. Particle fields, which are normally defined on four-dimensional real space-time, were defined on a complex four-dimensional space-time X^μ where the four apparent imaginary dimensions were assumed to be a vector quantum field $Y^\mu(y)$:

---

[15] The name Two-Tier is based on an analogy with a dynamic case where there are two tiers of activity – the low tier has a harmonic oscillator motion while the high tier imposes a further harmonic oscillator motion on it. See Blaha (2002).

[16] S. Blaha (2002), and (2012b): *Standard Model Symmetries, And Four And Sixteen Dimension Complex Relativity; The Origin Of Higgs Mass Terms (2012).* Chapters 5 and 6 below were abstracted from these books.

[17] The resulting perturbation theory has no infinities to any perturbative order. This finiteness requirement cannot be met by other renormalization mechanisms. Two-Tier Theory was developed by the author in 2001-2002.See chapters 5 and 6 for details. Since Two-Tier Theory gives essentially identical results to conventional perturbation theory at lower energies it is manifestly unitary in that sector. At ultra-high energies it will deviate from conventional perturbation theory results. However, it can be made unitary by normalizing the probabilities. A somewhat similar issue arises in Breit-Wigner fits to particle decay – not explicitly unitary due to the decay to other particles – thus physically acceptable. The ability of Two-Tier Theory to give finite perturbation theory results in universes of any dimension greater than or equal to four, makes Physics in other universes feasible giving a "universe friendly" Cosmos.

$$X^\mu(y) = y^\mu + i\, Y^\mu(y)/M_c^2 \quad (4.1)$$

where M_c was a large mass, perhaps, much beyond the order of the Planck mass, and the coordinates y^μ were those of a "subspace." The $Y^\mu(y)$ field was a gauge field similar to the electromagnetic field (but *not* the electromagnetic field) $Y^\mu(y)$ was a function of the subspace y coordinates. The real part of the space-time dimensions will be taken to be the subspace's y coordinates.[18]

The X^μ field can be viewed as a oscillating field built on base field coordinates y^μ. Because $Y^\mu(y)$ is a gauge field it introduces 2 degrees of freedom – dimensions into a quantum field theory based on it. Eq. 4.1 shows that the additional degrees of freedom are clearly of an "imaginary" nature. The implications of second quantization using X^μ are carefully explored in chapters 5 and 6.

A Two-Tier free scalar quantum field propagator has the form (chapter 5):

$$\phi(X) = \int d^3p\, N_m(p)\, [a(p) :e^{-ip\cdot(y + iY/M_c^2)}: + a^\dagger(p) :e^{ip\cdot(y + iY/M_c^2)}:] \quad (5.68)$$

4.1.1 Behavior of Feynman Propagators (From chapter 5)

The behavior of scattering amplitudes and probabilities at large and small distances is indicated by the features of a Feynman propagator. In chapter 5 we see (using chapter 5 equation numbers):

$$i\Delta_F^{TT}(y_1 - y_2) = i \int d^4p\, e^{-ip\cdot(y_1 - y_2)}\, R(\mathbf{p}, y_1 - y_2)/[(2\pi)^4(p^2 - m^2 + i\varepsilon)] \quad (5.77)$$

where

$$R(\mathbf{p}, y_1 - y_2) = \exp[-p^ip^j\Delta_{Tij}(y_1 - y_2)/M_c^4] \quad (5.78)$$

$$= \exp\{ -p^2[A(v) + B(v)\cos^2\theta]/[4\pi^2M_c^4z^2]\} \quad (5.79)$$

with

$$\Delta_{Tij}(z) = \int d^3k\, e^{-ik\cdot z}\, (\delta_{ij} - k_ik_j/\mathbf{k}^2)/[(2\pi)^3 2\omega_k] \quad (5.74)$$

where "T" indicates "Two-Tier".

$$z^\mu = y_1^\mu - y_2^\mu \quad (5.80)$$
$$z = |\mathbf{z}| = |\mathbf{y_1} - \mathbf{y_2}| \quad (5.81)$$
$$p = |\mathbf{p}| \quad (5.82)$$
$$v = |z^0|/z \quad (5.83)$$

$$A(v) = (1 - v^2)^{-1} + .5v \ln[(v - 1)/(v + 1)] \quad (5.84)$$

---

[18] The Two-Tier Theory can be directly generalized to duplex and quadplex fermions using $X^\mu(y) = y^\mu + i\, Y_y^\mu(y)/M_c^2$, and $X^\mu(z) = z^\mu + i\, Y_z^\mu(z)/M_c^2$, $X^\mu(u) = u^\mu + i\, Y_u^\mu(u)/M_c^2$, $X^\mu(v) = v^\mu + i\, Y_v^\mu(v)/M_c^2$ where the four $Y_i^\mu(y)$ commute with each other for i = y, z, u, v.

$$B(v) = v^2(1 - v^2)^{-1} - 1.5v \ln[(v - 1)/(v + 1)] \quad (5.85)$$

$$\mathbf{p\cdot z} = pz \cos\theta \quad (5.86)$$

and $|\mathbf{p}|$ denoting the length of a spatial vector **p** while $|z^0|$ is the absolute value of z^0.

The Gaussian damping factor R(p, z) for large momentum p is the same for both the positive and negative frequency parts of the Two-Tier Feynman propagator. It is also important to note that R(p, z) does not depend on p^0 (in the Y Coulomb gauge) and thus the integration over p^0 proceeds in the usual way to produce time-ordered positive and negative frequency parts.

4.1.2 Large Distance Behavior of Two-Tier Propagation

The large distance behavior of the Two-Tier Feynman propagator approaches the behavior of the conventional Feynman propagator since

$$R(\mathbf{p}, y_1 - y_2) \to 1 \quad (5.87)$$

when $(y_1 - y_2)^2$ becomes much larger than M_c^{-2}. *Thus the behavior of a conventional quantum field theory naturally emerges at large distance.* We will see that the conventional Standard Model is the large distance limit of the Two-Tier Standard Model thus *realizing a form of Correspondence Principle for Quantum Field Theory.* Some features of the conventional Standard Model that depend specifically on the existence of divergences, such as the axial anomaly, will be different in the Two-Tier Standard Model since it is a divergence-free theory.

4.2 Short Distance Behavior of Two-Tier Theories (from Chapter 5)

At short distances the Gaussian factor dominates, and radically changes, the behavior of the Feynman propagator eliminating its short distance singular behavior, and thus paving the way to finite quantum field theories. Near the light cone, $M_c^{-2} \gg -(y_1 - y_2)^2 \to 0$, we find the approximate propagator from eq. 5.77 to be

$$i\Delta_F^{TT}(y_1 - y_2) \approx \int d^3p\, [N(p)]^2 R(\mathbf{p}, y_1 - y_2) \quad (5.88)$$

since $e^{-ip\cdot(y_1 - y_2)}$ is approximately unity for small $(y_1 - y_2)$. We assume the mass of the ϕ particle is zero or is negligible at high energies so we set m = 0 to study the high energy behavior of eq. 5.88. Upon performing the integrations in eq. 5.88 for space-like $(y_1 - y_2)^2$ (and analytically continuing to the time-like regions[19,20]) we find the Feynman propagator:

$$i\Delta_F^{TT}(y_1 - y_2) \approx [z^2M_c^4/(4i\sqrt{A}\sqrt{B})] \ln[(\sqrt{A} + i\sqrt{B})/(\sqrt{A} - i\sqrt{B})] \quad (5.89)$$

---

[19] See S. Blaha, "Relativistic Bound State Models with Quasi-Free Constituent Motion", Phys. Rev. **D12**, 3921 (1975) and references therein.

[20] It should be noted that A and B in eqs. 5.84-85 have the same sign for $0 \le v < 1.1243$ thus making for easy analytic continuation across the light cone.

with A and B defined in eqs. 5.84 and 5.85. As $(y_1 - y_2)^2 \to 0$ from the space-like or time-like side of the light cone we find eq. 5.89 becomes (with $v \to 1$):

$$i\Delta_F^{TT}(y_1 - y_2) \to \pi M_c^4 |(y_1 - y_2)^\mu (y_1 - y_2)_\mu| /8 \qquad (5.90)$$

Eq. 5.90 has several noteworthy points:

1. The propagator is well behaved on the light cone and approaches zero smoothly from both space-like and time-like directions. In contrast, the *conventional* scalar Feynman propagator diverges as

$$i\Delta_F^{TT}(y_1 - y_2) \to [(y_1 - y_2)^\mu (y_1 - y_2)_\mu]^{-2} \quad \text{Conventional Result} \qquad (5.90A)$$

The good behavior of eq. 5.90 near the light cone will be seen later for other particle propagators with the net result that the usual infinities found in conventional quantum field theory are absent in Two-Tier quantum field theories.

2. The quadratic behavior *in coordinate space* of the propagator at short distances is equivalent to a high-energy behavior of

$$M_c^4 p^{-6} \qquad (5.91)$$

in momentum space. Thus we get the equivalent *of a higher derivative theory* in Two-Tier quantum field theory at high energies while retaining a positive definite energy spectrum. The problems of negative metric states that have plagued conventional higher derivative quantum field theories are avoided.

We conclude the low energy behavior of quantum field theories using Two-Tier coordinates is the same as conventional quantum field theories to good approximation. Thus unitarity is achieved to good approximation.[21]

We also conclude the *ultra-high* energy behavior of quantum field theories using Two-Tier coordinates avoids divergences found in conventional quantum field theories. Unitarity, which is not guaranteed, may be achieved by normalizing probabilities.

An absence of unitarity, might reflect a transition to a 6 dimension space-time with four real coordinates and two imaginary coordinates. It represents a form of leakage that might be compared to the Breit-Wigner decay of wave functions where probability conservation (unitarity) is absent due to particle decay.

---

[21] We note that unitarity, while a feature of conventional quantum field theories, has never been fully verified *experimentally*.

4.3 Imaginary Coordinates and Green's Functions

The difference between the Two-Tier propagator in eq. 5.89 above and the conventional propagator at ultra-high energies much above M_c. must be attributed to the extra pair of imaginary dimensions implicit in X^μ in eq. 4.1 due to the extra pair of dimensions of Y^μ. The extra dimensions cause the Feynman propagator to have an $R(\mathbf{p}, y_1 - y_2)$ factor in the integrand in eq. 5.77. One may represent the impact of this factor on the ultra-high energy form of the Feynman propagator in eq. 5.89 with an expression such as

$$R(\mathbf{p}, y_1 - y_2) \approx \int d^2q\, e^{-iq_1 z - iq_2 v} f(q_1, q_2) \qquad (4.2)$$

giving

$$i\Delta_F^{TT}(y_1 - y_2) \approx i \int d^4p\, d^2q\, e^{-ip\cdot(y_1 - y_2)}\, e^{-iq_1 z - iq_2 v}\, f(q_1, q_2)/[(2\pi)^4(p^2 + i\varepsilon)] \qquad (4.3)$$

due to the dependence of eq. 5.89 solely on z and v.

The Two-Tier theory generates an additional imaginary two dimensions based on the Y^μ field in eq. 4.1. These dimensions open the possibility of superluminal motion in ultra-high energy particle interactions and, thereby, the possibility of faster than light travel.

4.4 Ultra-High Energies, More Dimensions, and Duplex Wave Functions

It appears possible to increase the number of added dimensions to three by making the Y^μ field massive.

It is also possible to increase the number of added dimensions to four by making the Y^μ field massive without imposing a condition such as the Lorentz condition on the fields. In this case one would have four real dimensions and an additional four imaginary dimensions resulting in an effective duplex quantum theory as described earlier by the author.

4.5 Dynamics of Complex Forces

The complex form of coordinates of eq. 4.1 raises the possibility of circumventing the barrier at v = c that prevents transitions from sublight to superluminal motion. If one could achieve superluminal velocities then rapid journeys to the stars becomes feasible. To that end we define a Lorentz gauge $Y^0(y) = 0$ that gives the coordinate system:

$$X^i(y) = y^i + i\, Y^i(y)/M_c^2 \qquad (4.4)$$
$$X^0(y) = y^0 = t \qquad (4.5)$$

Then one may use the complex Newton's law

$$F_{real}^{\ i} = dp_{real}^{\ i}/dt \qquad (4.6)$$
$$F_{imag}^{\ i} = dp_{imag}(Y^i/M_c^2, t)/dt \qquad (4.7)$$

where $F_{real}^{\ i}$ and $F_{imag}^{\ i}$ are real and imaginary forces, and

$$y_{real}{}^{i} = y^{i} \qquad (4.8)$$
$$y_{imag}{}^{i} = Y^{i}(y, t)/M_c{}^{2} = Y^{i}(y_{real}, t)/M_c{}^{2} \qquad (4.9)$$

4.5.1 Simple Complex Velocity Model

This model uses a complex-*valued* force to drive the acceleration of a particle. As the particle velocity increases its imaginary part, which had been negligible, becomes much greater. As a result the barrier at v = c is passed. The energy and momentum remains finite – as does $\gamma = (1 - v^2/c^2)^{-½}$.

The real-valued coordinates are

$$y_{real}{}^{i} = y^{i} \qquad (4.10)$$

We will assume the imaginary part of the coordinates has the same direction as the real part of the coordinates. As a result

$$y_{imag}{}^{i} = Y^{i}(\mathbf{y}_{real}, t)/M_c{}^{2} \qquad (4.11)$$

We now assume there is only one spatial coordinate. We further assume the field is a Fourier component of an electromagnetic-like field:

$$\begin{aligned} y_{imag} &= Y(\mathbf{y}_{real}, t)/M_c{}^{2} \qquad (4.12) \\ &= \sin(\omega(t - y_{real}))/M_c{}^{2} \end{aligned}$$

resulting in

$$v_{imag} = \omega(1 - dy_{real}/dt)\cos(\omega(t - y_{real}))/M_c{}^{2} \qquad (4.13)$$

Assuming a constant force F we find the complex momentum

$$p = Ft$$

starting at p = 0 for t = 0. We set the momentum:

$$\begin{aligned} p &= mc\gamma(v_{real} + iv_{imag}) \qquad (4.14) \\ &= mc(v_{real} + iv_{imag})/(1 - v_{real}{}^{2} + v_{imag}{}^{2})/c^2)^{½} \\ &= p_{real} + ip_{imag} \\ &= mc\gamma v_{real} + imc\gamma v_{imag} = Ft \qquad (4.14a) \end{aligned}$$

using

$$\gamma = 1/(1 - (v_{real}{}^{2} + v_{imag}{}^{2})/c^2))^{½} \qquad (4.15)$$

and assuming the metric

$$ds^2 = ds_{real}{}^{2} - ds_{imag}{}^{2} \qquad (4.16)$$

with signatures (1, –1, –1, –1, –1).

At v_{real} = c the momentum p is finite and the barrier to superluminal can be crossed.

Eq. 4.14a implies

$$v_{real} = v_{real}(t, v_{imag}) \tag{4.17}$$

and together with eq. 4.13 implies

$$v_{imag} = v_{imag}(t, y_{real}, v_{real}) \tag{4.18}$$
$$= \omega(1 - v_{real})\cos(\omega(t - y_{real}))/M_c^2 \tag{4.19}$$

Eq. 4.19 reveals v_{imag} is an oscillating function of time and y_{real}. Thus the dynamics can be set up to avoid the v = c barrier. See Figs. 4.1 – 4.3.

4.5.2 Oscillating Imaginary Velocity

The oscillating velocity v_{imag} in eq. 4.19 enables the barrier at v = c to be overcome. The oscillation in v_{imag} prevents a great rise in its magnitude.

The dependence of the solution of eq. 4.17 on v_{imag} is mild. *Consequently v_{real} will be increasing "steadily" with time, while undergoing a relatively small oscillation. The overall distance y_{real} will be increasing with time and support faster than light motion. The net imaginary distance y_{imag} will be oscillating with time and will not be growing in magnitude.*

The real and imaginary behavior of this model's starship motion seems reasonable and would support travel to the stars.

The Superluminal motion in this model is directly based on Two-Tier Theory coordinates.

4.5.3 Force for Imaginary Velocity

The origin of the forces is the issue. A simple engine design that might give complex forces that would enable starships to penetrate to superluminal velocities appears in Figs. 4.1 - 4.3. We will consider more detailed models in a future book.[22]

---

[22] Some of the author's previous more detailed books include: (2014b) *All the Megaverse! Starships Exploring the Endless Universes of the Cosmos Using the Baryonic Force;*
(2014c) *All the Megaverse! II Between Megaverse Universes: Quantum Entanglement Explained by the Megaverse Coherent Baryonic Radiation Devices – PHASERs Neutron Star Megaverse Slingshot Dynamics Spiritual and UFO Events, and the Megaverse Microscopic Entry into the Megaverse.*

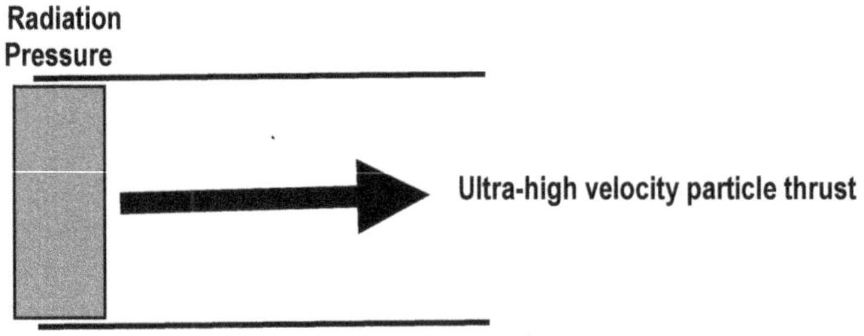

Figure 4.1. A simple design for a "rocket" engine that may generate a thrust of particles in complex coordinates motion. A ultra-large burst of radiation pressure makes a mass of particles accelerate in complex motion to a speed beyond c. The complex velocity of the particles, generating increasing starship real velocity, transcends v = c enabling faster than light starship motion. The acceleration process may be in a pulsed form that causes a starship to accelerate "gently" to avoid crushing human occupants of the ship. The remainder of the ship is not pictured.

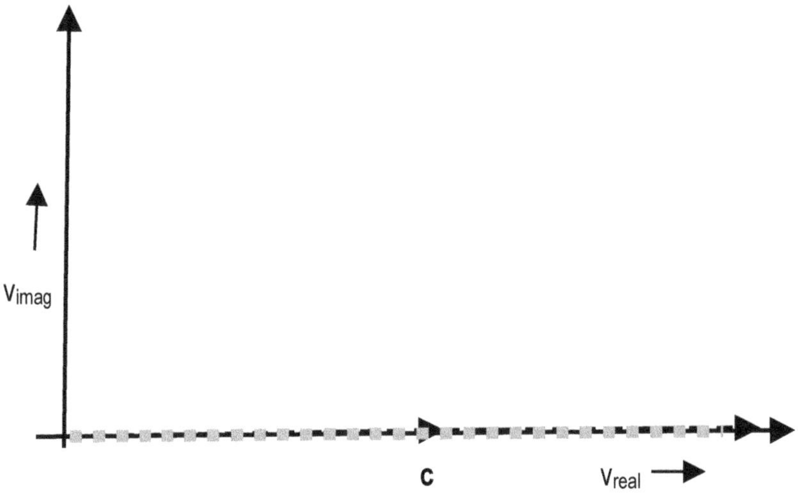

Figure 4.2. Plot of the complex starship velocity as a function of time. The starship gains increasing real velocity (solid line) and yet evades the singularity at v_{real} = c. The imaginary part of its velocity oscillates (gray line). The real velocity v_{real} has a mild oscillation as it increases with time (not shown). The imaginary velocity v_{imag} oscillates with time.

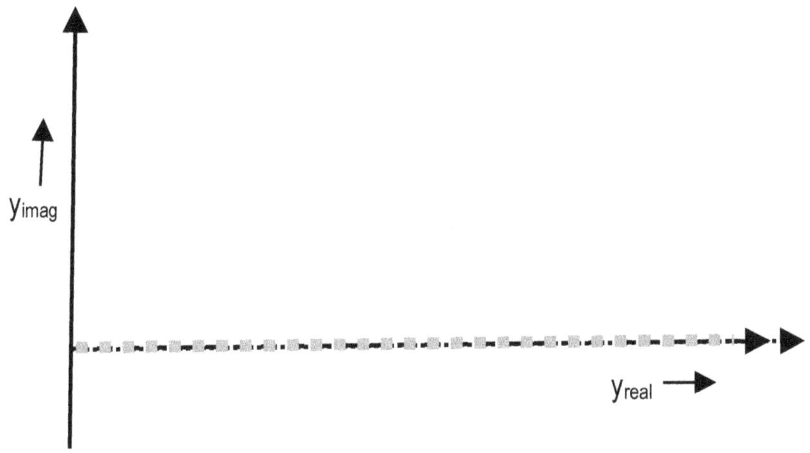

Figure 4.3. Plot of the complex starship movement as a function of time. The starship has increasing real distance (dashed line). The starship has oscillating imaginary distance (gray line). The oscillations are not shown in the figure.

4.6 Superluminal Starships

The above analysis is based on the form of complex coordinates that emerges when velocities become extremely high. At low velocities their imaginary part is oscillating and negligible. We now consider some issues raised involving superluminal starship travel.

4.6.1 The Need for Speedy Transit

Travel to the stars as envisioned in most proposals would take many years or decades. This travel time more or less ruins the possibility of significant exploration and colonization. The ultra-high velocity approach, that we suggest should be explored, offers the possibility of "mass" migration to the stars rather like the migration to America over the past centuries. Then Humanity would enjoy a refreshing expansion similar to the upsurge in Europe after America's discovery.

4.6.2 Potential Problems

There are problems that arise in superluminal starship travel:

1. High acceleration required for acceleration to ultra-high speeds. The human body cannot take greater than 10 g's. Some form of hibernation during transit will undoubtedly be needed.

2. The engine will need new materials due to the likelihood of erosion of the starship "tail" assembly and other parts during transit.

Chapter 9 of Blaha (2014b) provides a detailed description of issues. Chapter 8 raises the possibility of exit from our universe into the Megaverse. Chapter 10 describes a scenario for design and development of a starship.

4.6.3 Interstellar Communications

Given the great distances involved the instantaneous nature of Quantum Entanglement recommends itself. There could be transmitter/receivers on the starship and Earth for instantaneous communication. The entanglement at the initial point of the voyage should be for entangled bytes of spins (eventually multibytes of spins) as we pictured in section 10.2 of Blaha (2014b). For example, one may start initially with one byte (8 entangled spins) of Morse code. It is easy to envision arrays of spins that could transmit kilobytes or even megabytes of data, and voice transmissions, instantaneously in bursts.

5. Two-Tier Quantum Field Theory and the Flatverse

In a series of books culminating in Blaha (2005a) we developed a new approach to quantum field theory (Two-Tier Quantum Field Theory) that eliminated the issues of infinities in renormalizations that plagued the traditional form of quantum field theory since the 1940's. The key feature of this approach was the appearance of a Gaussian damping factor in each propagator of each particle in a perturbation theory calculation. Consequently all Stueckelberg-Feynman diagrams to any order yielded a finite result.

A significant aspect of this approach was that for small momenta the usual results of The Standard Model and QED were obtained. For very large momenta of the order of a certain mass (perhaps the Planck mass) perturbation theory calculations were vastly different from the lower momenta results, and all diagrams had finite values without cutoffs or other artifices.

The key to this approach was the use of two sets of coordinates. One set of coordinates were ordinary c-value coordinates. The other set of coordinates defined in terms of the first set were q-number coordinates that were effectively quantum "smeared". The smearing led directly to Stueckelberg-Feynman propagators with a Gaussian factor that suppressed potential divergences for large values of the integration momenta. Yet, happily, the low momenta approximations to the diagrams were of the usual form and thus led to the usual results in QED and Standard Model calculations.

In this chapter we associate the c-number coordinates in quantum field theory to flatverse coordinates, and the q-number coordinates to the coordinates of our universe. Thus we use the framework established in preceding chapters to provide a deeper basis for Two-Tier Quantum Field Theory. One important consequence of this development is to establish a unity in our understanding of space-time, symmetries, the Standard Model and Gravitation. Much of the following material in this chapter is abstracted from Blaha (2005a).

Chapter 6 describes Two-Tier perturbation theory.

5.1 Relation of Two-tier Coordinates to Coordinates of the Flatverse and Our Universe

In Blaha (2005a) and earlier books we considered a coordinate system labeled X^μ that had a real-valued c-number part and an imaginary-valued q-number part. Particle fields, which are normally defined on four-dimensional real space-time, were defined on a complex four-dimensional space-time X^μ where the four imaginary dimensions were assumed to be a vector quantum field $Y^\mu(y)$:

$$X^\mu(y) = y^\mu + i\, Y^\mu(y)/M_c^2 \tag{5.1}$$

where M_c was a large mass, perhaps, of the order of the Planck mass, and the coordinates y^μ were those of a "subspace." The $Y^\mu(y)$ field was a gauge field similar to the electromagnetic field (but *not* the electromagnetic field) $Y^\mu(y)$ was a function of the subspace y coordinates. The real part of the space-time dimensions will be taken to be the subspace's y coordinates.

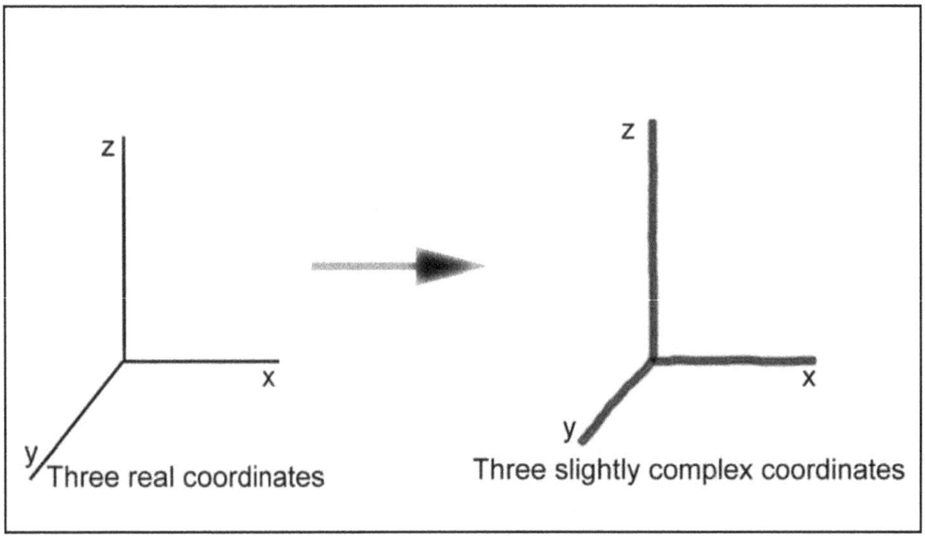

Figure 5.1. The change from purely real space to a slightly complex space with imaginary quantum fluctuations for each spatial axis in the Coulomb gauge of the Y field.

Based on our development of a space-time theory in which our universe is a complex 4-dimensional surface embedded in a 16-dimensional complex space, we now take the view that y^μ is a real-valued set of coordinates in the flatverse (made real by a Reality group transformation if necessary). The flatverse coordinates can be restricted to 4 dimensions in the case where our universe is assumed to be flat. We make that assumption in Blaha (2005a).

$X^\mu(y)$ are the coordinates of a point in our universe. These coordinates are related to y^μ - by an equation such as eq. 2.1 but with the added complication of having a q-number part. The proper view of the X^μ coordinates is that the c-number part of their value is real, or made real by a Reality group transformation. The q-number part is not subject to the Reality group transformation. The discussion of Blaha (2005a) was based on the assumption of a flat space-time in our universe. This discussion is applicable to particle physics in our part of the universe because the curvature of the universe is almost flat, and long-range interactions such as electromagnetism and gravitation are well accounted for at long range by classical theory.

Thus our 16-dimensional flatverse construction can accommodate Two-Tier Quantum Field Theory under the assumption that our universe is flat. The extension of the Two-Tier theory to a curved universe does not appear to offer any difficulties in principle but does involve more detail and 16-dimensional y coordinates with their accompanying changes in integrations.[23]

5.1.1 The q-number part of X^μ

The imaginary q-number part of space-time (which has not been experimentally seen) consists of quantum fluctuations of a massless vector quantum field that are suppressed further by a large mass scale – perhaps of the order of the Planck mass, thus reducing the imaginary part to the infinitesimal. The effects of the q-number part only become appreciable in quantum field theory at energies of the order of M_c. At these energies a divergence suppressing exponential Gaussian factor (seen later) in each particle (and ghost) propagator serves to make perturbation theory calculations ultra-violet finite – including calculations in Quantum Gravity.

The formalism that we developed introduced a new form of interaction that did not have the form of the simple polynomial interactions that have hitherto dominated quantum field theories. This form of interaction takes place via the composition of quantum fields – fields within fields.

5.2 Second Quantization Starting With a C-Number X^μ

We will begin by considering the case of a scalar quantum field theory. We assume a real underlying y subspace. Since X^μ is a set of coordinates, we choose to define a scalar field ϕ as a function of X^μ, which in turn is a function of the y^ν coordinates. We will provisionally second quantize ϕ treating the X^μ coordinates as c-number coordinates using a conventional approach.

The momentum conjugate to ϕ is:

$$\pi_\phi = \partial L_F/\partial\phi' \equiv \partial L_F/\partial(\partial\phi/\partial X^0) \tag{5.2}$$

Following the canonical quantization procedure, π and ϕ become Hermitian operators with equal time ($X^0 = X^{0\prime}$) commutation rules:

$$[\phi(X), \phi(X')] = [\pi_\phi(X), \pi_\phi(X')] = 0 \tag{5.3}$$

$$[\pi_\phi(X), \phi(X')] = -i\,\delta^3(\mathbf{X} - \mathbf{X'}) \tag{5.4}$$

The Hamiltonian is defined by

---

[23] The larger number of dimensions does not create renormalization issues because of the Gaussian exponential factors that appear in all perturbative calculations. These factors eliminate divergences due to a larger number of dimensions.

$$H_F = P_{F0} = \int d^3X\ T_{F00} \tag{5.5}$$

The standard Fourier expansion of the solution to the Klein-Gordon equation

$$(\Box + m^2)\phi(X) = 0 \tag{5.6}$$

where

$$\Box = \partial/\partial X^{\nu}\,\partial/\partial X_{\nu} \tag{5.7}$$

is

$$\phi(X) = \int d^3p\ N_m(p)\ [a(p)\ e^{-ip.X} + a^{\dagger}(p)\ e^{ip.X}] \tag{5.8}$$

where

$$N_m(p) = [(2\pi)^3 2\omega_p]^{-1/2} \tag{5.9}$$

and

$$\omega_p = (\mathbf{p}^2 + m^2)^{1/2} \tag{5.10}$$

The commutation relations of the Fourier coefficient operators are:

$$[a(p), a^{\dagger}(p')] = \delta^3(\mathbf{p} - \mathbf{p'}) \tag{5.11}$$

$$[a^{\dagger}(p), a^{\dagger}(p')] = [a(p), a(p')] = 0 \tag{5.12}$$

The reader will recognize the quantization procedure is formally identical to the standard canonical quantization procedure of a free scalar quantum field.

In the case of spin ½, spin 1 and spin 2 fields the standard quantization procedure *in terms of the X coordinate system* can also be followed in a way similar to the procedure in standard texts.

The quantum field theory developments in this chapter may disturb some readers since we are combining operators with Dirac delta functions and using other unusual operator expressions. These concerns are put at rest when we realize path integral formulations give precisely the same results as the present development. See Blaha (2005a).

5.3 X^{μ} Coordinate Quantization

In this section we quantize the coordinates X^{μ} as a vector field defined on a fundamental c-number coordinate system y^{ν} of the same dimensionality. We will assume the y^{ν} space is a "normal" flat Minkowski space with three spatial and one time dimensions. *This space is the flatverse described in earlier chapters but is only required to be 4-dimensional since we assume our universe to be flat.* A generalization to 16-dimensional flatverse space is straightforward but not necessary for perturbation theory calculations in particle physics except near Black Hole horizons or anomalously curved space-time regions, such as worm holes, that have not been observed.

Thus we will assume X^{μ} has three spatial dimensions and one time dimension. For reasons primarily of simplicity (primarily to avoid multiple time coordinates) we will assume the X^μ fields are similar to the free electromagnetic vector potential A^μ with the Lagrangian:

$$\mathscr{L}_C = +\tfrac{1}{4}\, M_c^4 F^{\mu\nu} F_{\mu\nu} \tag{5.13}$$

$$F_{\mu\nu} = \partial X_\mu/\partial y^\nu - \partial X_\nu/\partial y^\mu \tag{5.14}$$

where M_c is a mass scale raised to the fourth power that is required on dimensional grounds and serves to set the scale for new Physics as we will see later. *Note the sign in eq. 5.13 is not negative – superficially contrary to the conventional electromagnetic Lagrangian. The reason for this difference is that the field part of X^μ is imaginary.* Thus $\mathscr{L}_C$ winds up having the correct sign after taking account of the factors of i in the field strength $F_{\mu\nu}$.

We assume X^μ is complex[24] with the form:

$$X_\mu(y) = y_\mu + i\, Y_\mu(y)/M_c^2 \tag{5.15}$$

where $Y_\mu(y)$ is a quantum field, and y_μ is a real-valued, c-number 4-vector (the flatverse coordinates restricted to 4 dimensions). If X^μ has this form, then we can define

$$F_{\mu\nu} = i\, (\partial Y_\mu/\partial y^\nu - \partial Y_\nu/\partial y^\mu)/M_c^2 \tag{5.16}$$

and

$$F_{Y\mu\nu} = \partial Y_\mu/\partial y^\nu - \partial Y_\nu/\partial y^\mu \tag{5.17}$$

The Lagrangian has the form of the conventional electromagnetic Lagrangian:

$$\mathscr{L}_C = -\tfrac{1}{4}\, F_Y^{\mu\nu} F_{Y\mu\nu} \tag{5.18}$$

This Lagrangian is used to generate field equations and develop a canonical quantization that is completely analogous to Quantum Electrodynamics.

5.3.1 Gauge Invariance

The gauge invariance of the Lagrangian allows us to choose a convenient gauge. The gauge invariance of the full Lagrangian

$$\mathscr{L}_s = \mathscr{L}_F(\phi(X), \partial\phi/\partial X^\mu)\, J + \mathscr{L}_C(X^\mu(y), \partial X^\mu(y)/\partial y^\nu) \tag{5.19}$$

---

[24] Theories of quantum mechanics, and quantum fields, in complex and quaternion spaces have been considered by numerous authors. For example see C. M. Bender, D. C. Brody and H. F. Jones, "Complex Extension of Quantum Mechanics" Phys. Rev. Letters **89**, 270401-1 (2002) and references therein; S. L. Adler and A. C. Millard, "Generalized Quantum Dynamics as Pre-Quantum Mechanics", Princeton Univ. preprint arXiv:hep-th/9508076 (1995) and references therein. These theories are all very different from the theory presented herein.

is based on the gauge invariance of $\mathscr{L}_C$, and the invariance of $J\mathscr{L}_F$ in the form of translational invariance

$$X^\mu(y) \rightarrow X^\mu(y) + \delta X^\mu(y) \tag{5.20}$$

for the case of a translation of X of the form of a gauge transformation:

$$\delta X^\mu(y) = \partial\Lambda(y)/\partial y_\mu$$

In this case we find[25]

$$\int d^4y\ \Lambda(y)\ \partial\ [\ J\ \partial/\partial X^\mu\ \mathscr{T}_{F\mu\nu}\ ]/\partial y_\nu\ = 0$$

after a partial integration and so we obtain the differential conservation law:

$$\partial\ [J\ \partial/\partial X^\mu\ \mathscr{T}_{F\mu\nu}\ ]/\partial y_\nu\ = 0 \tag{5.21}$$

since $\Lambda(y)$ is arbitrary. This conservation law is trivially obeyed since

$$\partial/\partial X^\mu\ \mathscr{T}_{F\mu\nu}\ = 0 \tag{5.22}$$

Thus translational invariance in the $\mathscr{L}_F$ sector together with standard gauge invariance in the $\mathscr{L}_C$ sector automatically guarantees the Y field gauge invariance of the total Lagrangian. Basically we use the separate invariance of each term of

$$L = \int d^4y\ [\mathscr{L}_F\ J + \mathscr{L}_C\ ] = \int d^4X\ \mathscr{L}_F\ + \int d^4y\ \mathscr{L}_C = L_F + L_C \tag{5.23}$$

under a constant translation $X^\mu \rightarrow X^\mu + \delta X^\mu$ where δX^μ is constant to establish eq. 5.22. Then we consider a position dependent translation/gauge transformation, which taken together with eq. 5.22, establishes invariance under the position dependent translation/gauge transformation eq. 5.20.

Having established invariance under gauge transformations we now choose to use the most convenient gauge – the Coulomb gauge[26]:

$$\partial Y^i/\partial y^i = 0 \tag{5.24}$$

which, in the absence of external sources, allows us to set

$$Y^0 = 0 \tag{5.25}$$

---

[25] Eq. A.106 of Blaha (2005a).

[26] It is also possible to quantize using an indefinite metric that preserves manifest Lorentz covariance as was done by Gupta and Bleuler for the electromagnetic field.

since Y^0 does not have a canonically conjugate momentum. A conventional treatment leads to the equal time commutation relations:

$$[Y^\mu(\mathbf{y}, y^0), Y^\nu(\mathbf{y}', y^0)] = [\pi^\mu(\mathbf{y}, y^0), \pi^\nu(\mathbf{y}', y^0)] = 0 \tag{5.26}$$

$$[\pi^j(\mathbf{y}, y^0), Y_k(\mathbf{y}', y^0)] = -i\,\delta^{tr}{}_{jk}(\mathbf{y} - \mathbf{y}') \tag{5.27}$$

where

$$\pi^k = \partial\mathcal{L}_C / \partial Y_k' \tag{5.28}$$

$$\pi^0 = 0 \tag{5.29}$$

$$\delta^{tr}{}_{jk}(\mathbf{y} - \mathbf{y}') = \int d^3k\, e^{i\,\mathbf{k}\cdot(\mathbf{y} - \mathbf{y}')}(\delta_{jk} - k_j k_k/\mathbf{k}^2)/(2\pi)^3 \tag{5.30}$$

$$Y_k' = \partial Y_k/\partial y^0 \tag{5.31}$$

The Coulomb gauge reveals the two degrees of freedom that are present in the vector potential. The Fourier expansion[27] of the vector potential is:

$$Y^i(y) = \int d^3k\, N_0(k) \sum_{\lambda=1}^{2} \varepsilon^i(k, \lambda)[a(k,\lambda)\, e^{-ik\cdot y} + a^\dagger(k,\lambda)\, e^{ik\cdot y}] \tag{5.32}$$

where

$$N_0(k) = [(2\pi)^3 2\omega_k]^{-1/2} \tag{5.33}$$

and (since m = 0)

$$\omega_k = (\mathbf{k}^2)^{1/2} = k^0 \tag{5.34}$$

with $\vec{\epsilon}(k, \lambda)$ being the polarization unit vectors for $\lambda = 1,2$ and $k^\mu k_\mu = 0$.

The commutation relations of the Fourier coefficient operators are:

$$[a(k,\lambda), a^\dagger(k',\lambda')] = \delta_{\lambda\lambda'}\delta^3(\mathbf{k} - \mathbf{k}') \tag{5.35}$$

$$[a^\dagger(k,\lambda), a^\dagger(k',\lambda')] = [a(k,\lambda), a(k',\lambda')] = 0 \tag{5.36}$$

and the polarization vectors satisfy

$$\sum_{\lambda=1}^{2} \varepsilon_i(k, \lambda)\varepsilon_j(k, \lambda) = (\delta_{ij} - k_i k_j/\mathbf{k}^2) \tag{5.37}$$

It will be convenient to divide the Y field into positive and negative frequency parts:

---

[27] The Fourier expansion of the $Y^i(y)$

$$Y^{+}_{i}(y) = \int d^3k\ N_0(k) \sum_{\lambda=1}^{2} \varepsilon_i(k, \lambda)\ a(k,\lambda)\ e^{-ik\cdot y} \qquad (5.38)$$

and

$$Y^{-}_{i}(y) = \int d^3k\ N_0(k) \sum_{\lambda=1}^{2} \varepsilon_i(k, \lambda)\ a^{\dagger}(k,\lambda)\ e^{ik\cdot y} \qquad (5.39)$$

For later use we note the commutator between the positive and negative frequency parts is:

$$[\, Y^{-}_{j}(y_1), Y^{+}_{k}(y_2)] = -\int d^3k\ e^{ik\cdot(y_1 - y_2)}\ (\delta_{jk} - k_jk_k/\mathbf{k}^2)/[(2\pi)^3 2\omega_k] \quad (5.40)$$

5.4 Bare ϕ Particle States

We now turn to consider the ϕ particle discussed earlier and its states. Creation and annihilation operators can be used to define "bare" free particle states. Bare free particle states are states that are not dressed with coherent states of Y quanta. For example a bare one-particle state of momentum p is

$$|p\rangle = a^{\dagger}(p)|0_{\phi}\rangle \qquad (5.41)$$

with corresponding bare bra state

$$\langle p| = \langle 0_{\phi}|a(p) \qquad (5.42)$$

where the vacuum is defined as usual:

$$a(p)|0_{\phi}\rangle = 0 \qquad (5.43)$$

$$\langle 0_{\phi}|a^{\dagger}(p) = 0 \qquad (5.44)$$

Multi-particle bare states can also be defined in the conventional way with products of creation and annihilation operators applied to the vacuum.

5.5 Y Coordinate Coherent States

States can also be defines for the quantized Y field. These states will be similar in form to electromagnetic photon states but play a different role in our approach since they are in fact coordinate excitation states for the imaginary part of X^{μ}. Thus the scalar field (and other particle fields) will exist in our real four-dimensional space (and real flatverse space) with quantum excitations into imaginary quantum dimensions. These excitations become significant at high energies. At the low energies, with which we are familiar, space-time appears real. At very high energies space-time becomes slightly complex.

There are two types of imaginary coordinate excitations: 1.) Quantum excitations into Fock states consisting of superpositions of states with a definite finite number of Y "particles" and 2.) Imaginary coordinate excitations into *coherent* Y states with an "infinite" number of particles. Coherent states can be viewed as representing "classical" fields.

In this section we will consider Y field states with a definite number of excitations ("particles"). The creation and annihilation operators of the Y field can be used to define free particle states. For example a one particle state can be defined by

$$|k, \lambda> = a^{\dagger}(k, \lambda)|0_Y> \quad (5.45)$$

with corresponding bra state

$$<k, \lambda| = <0_Y|a(k, \lambda) \quad (5.46)$$

where the "coordinate vacuum" is defined as usual:

$$a(k, \lambda)|0_Y> = 0 \quad (5.47)$$
$$<0_Y|a^{\dagger}(k, \lambda) = 0 \quad (5.48)$$

Multi-particle states can also be defined in the conventional way as products of the creation and annihilation operators applied to the vacuum. The set of all states, each containing a finite number of "particles", constitutes a Fock space.

A state with a finite number of Y "particles" represents a quantum fluctuation into imaginary quantum dimensions. Such states do not appear in Two-Tier quantum field theory since the Y field is a free field and has no source. Thus they appear only as part of normal particles.

A normal particle, such as a ϕ particle, has a coherent state of Y quanta associated with it, which play a role in interactions. The Y coherent state part of a normal particle can be viewed as boring an infinitesimal "hole" into an extra pair of imaginary dimensions in a neighborhood of the particle of a radial extent set by the length M_c^{-1}.

Coherent Y states bring us closer what we might consider to be "classical" imaginary dimensions – dimensions that we can, in principle, experience as we do normal dimensions. Let us define the coherent state[28]

$$| y, p> = e^{-\mathbf{p}\cdot\mathbf{Y}^{-}(y)/M_c^2}|0_Y> \quad (5.49)$$

This state is an eigenstate of the coordinate operator $Y^{+}(y')$:

$$Y^{+}_j(y_1)|y_2, p> = -[Y^{+}_j(y_1), \mathbf{p}\cdot\mathbf{Y}^{-}(y_2)]/M_c^2|y, p> \quad (5.50)$$

[28] Coherent states are well known in the physics literature. See for example T. W. B. Kibble, J. Math. Phys. **9**, 315 (1968) and references therein; V. Chung, Phys. Rev. **140**, B1110 (1965); J. R. Klauder, J. McKenna, and E. J. Woods, J. Math. Phys. **7**, 822 (1966) and references therein.

$$= - \int d^3k\, [N_0(k)]^2\, e^{ik\cdot(y_2 - y_1)}\, (p_j - k_j\mathbf{p}\cdot\mathbf{k}/\mathbf{k}^2)/M_c^2\, |y, p> \quad (5.51)$$

$$= p^i\Delta_{Tij}(y_1 - y_2)/M_c^2\, |y, p> \quad (5.52)$$

where $p^i\Delta_{Tij}(y_1 - y_2)/M_c^2$ is the eigenvalue of $Y^+_j(y_1)$. As we will see in the next chapter, the eigenvalue of Y^+ becomes large as $(y_1 - y_2)^2 \to 0$. Thus the imaginary quantum dimensions become significant at very short distances, and significantly modify the high-energy behavior of quantum field theories. In particular, quantum dimensions have a significant effect when

$$(y_1 - y_2)^2 \lessapprox (4\pi^2 M_c^2)^{-1} \quad (5.53)$$

We assume the mass scale M_c is very large – perhaps of the order of the Planck mass (1.221×10^{19} GeV/c^2). Thus imaginary quantum dimensions are far from detectable in today's "low" energy experiments. Their effect is significant in the analysis of the first instants after the Big Bang.[29]

5.6 Generation of Quantum Dimensions by the ϕ (X) field

The ϕ(X) field generates quantum dimensions via coherent states from the vacuum:

$$\phi(X) = \int d^3p\, N_m(p)\, [a(p)\, e^{-ip\cdot(y + iY/M_c^2)} + a^\dagger(p)\, e^{ip\cdot(y + iY/M_c^2)}] \quad (5.54)$$

with the result

$$\phi(X)|0> = \int d^3p\, N_m(p)\, a^\dagger(p)\, e^{ip\cdot(y + iY/M_c^2)}|0> \quad (5.55)$$

is a superposition of coherent Y states plus one scalar particle. The vacuum state |0> is the product of the ϕ and Y vacuum states $|0> = |0_Y>|0_\phi>$. We will use |0> in most of the following discussions.

We can also define coherent Y states with total momentum q using the expression:

$$|q\, Y> = \int d^4y\, e^{iq\cdot X(y)}|0> = \int d^4y\, e^{iq\cdot(y + iY/M_c^2)}|0> \quad (5.56)$$

Expanding the Y part of the exponential in eq. 3.48 gives

$$|q\, Y> = \sum_{n=0}^{\infty}(-1)^n(n!)^{-1}\prod_{j=1}^{n}\left(\int d^3k_j N_0(k_j)\right)\delta^4\left(q - \sum_{s=1}^{n} k_s\right)\prod_{r=1}^{n}\sum_{\lambda_r=1}^{2} \mathbf{q}\cdot \varepsilon(k_r, \lambda_r)a^\dagger(k_r,\lambda_r)|0> \quad (5.57)$$

---

29 Blaha (2004).

which indicates that the sum of the Y particle momenta for each term in the expansion is q.

5.7 Hamiltonian and Momentum for Particle and Coordinate States

This section summarizes the results in Blaha (2005a). The Hamiltonian for the separable, coordinate quantized, scalar quantum field theory is:

$$H_s = \int d^3y\, \mathcal{H}_s \tag{5.58}$$

with

$$\mathcal{H}_s = J\mathcal{H}_F + \mathcal{H}_C \tag{5.59}$$

$$\mathcal{H}_F = ½\, [\pi_\phi^{\ 2} + (\partial\phi/\partial X^i)^2 + m^2\phi^2] \tag{5.60}$$

$\mathcal{H}_F$ is the conventional scalar particle Hamiltonian when viewed as a function of the X coordinates. $\mathcal{H}_C$ has the same form as the conventional electromagnetic Hamiltonian

$$\mathcal{H}_C = ½\, (E_Y^{\ 2} + B_Y^{\ 2}) \tag{5.61}$$

where

$$E_Y^{\ i} = -\partial Y^i/\partial y^0 \tag{5.62}$$

$$B_Y^{\ i} = \varepsilon^{ijk}\, \partial Y_j/\partial y^k \tag{5.63}$$

Using fourier expansions of ϕ and X^μ we obtain the following expression for the normal-ordered Hamiltonian H_S:

$$P_s^{\ 0} \equiv H_s = \int d^3y :H_s: \tag{5.64}$$

$$H_s = \int d^3p(\mathbf{p}^2 + m^2)^{½}a^\dagger(p)a(p) + \int d^3k \sum_{\lambda=1}^{2} (\mathbf{k}^2)^{½}\, a^\dagger(k, \lambda)a(k, \lambda) \tag{5.65}$$

where : : indicates normal ordering and where we perform a functional integration over X (Note the Jacobian is present within $\mathcal{H}_s$.) for the particle part of the Hamiltonian $\mathcal{H}_F$. The Hamiltonian is manifestly positive definite.

The spatial momentum is specified by

$$P_S^{\ j} = -\int d^3X :\pi_\phi(X)\partial\phi(X)/\partial X_j: + \int d^3y :E_Y^{\ i}\partial Y^i/\partial y_j: \tag{5.66}$$

$$= \int d^3p\, p^j\, a^\dagger(p)a(p) + \int d^3k \sum_{\lambda=1}^{2} k^j\, a^\dagger(k, \lambda)a(k, \lambda)$$

where the first term in eq. 5.66 follows from $\int d^3X$. The momentum operator generates displacements in ϕ

$$[P_s^{\mu}, \phi(X)] = -\, i\partial\phi/\partial X_\mu \qquad (5.67)$$

5.8 Normal Ordered Particle Fields Required

The Fourier expansion of ϕ does require one refinement – the exponential terms in X^μ must be *normal ordered* to avoid infinities in the unequal time commutation relations:

$$\phi(X) = \int d^3p\, N_m(p)\, [a(p) :e^{-ip\cdot(y + iY/M_c^{\,2})}: + a^\dagger(p) :e^{ip\cdot(y + iY/M_c^{\,2})}:] \qquad (5.68)$$

Since the Hamiltonian as well as other quantities are normal ordered in quantum field theory the additional requirement of normal ordering in the field operator is merely an extension of a standard procedure to a more complex situation and is not disturbing. The unequal time commutation relation of the normal ordered ϕ field is:

$$[\phi(X^\mu(y_1)), \phi(X^\mu(y_2))] = i\Delta(y_1 - y_2) + O(1/M_c^2) \qquad (5.69)$$

where

$$\Delta(y_1 - y_2) = -\, i \int d^3k\, (e^{-ik\cdot(y_1 - y_2)} - e^{ik\cdot(y_1 - y_2)})/[(2\pi)^3 2\omega_k] \qquad (5.70)$$

is a familiar c-number invariant singular function. The additional terms in eq. 5.69 are q-number terms that become significant at very short distances of the order M_c^{-1}. Thus precise measurements of field strengths at larger distances are limited by standard quantum effects as indicated by the commutation relation.

The principle of *microscopic causality* is violated at extremely short distances of the order M_c^{-1} since the commutator (eq. 5.69) is non-zero, in general, for space-like distances of the order of M_c^{-1} due to the q-number terms. This violation is not experimentally measurable now – and for the foreseeable future – and reflects a type of non-locality at extremely short distances.

The short distance behavior of Two-Tier quantum field theory leads to the elimination of divergences resulting in finite interacting quantum field theories.

5.9 Vacuum Fluctuations

While the expectation value of a *conventional* free scalar field $\phi_{conv}(X)$ is zero in a conventional quantum field theory:

$$<0|\phi_{conv}(X)|0> = 0 \qquad (5.71)$$

the vacuum fluctuations of *conventional* scalar quantum field theory are quadratically divergent:

$$<0|\phi_{conv}(X)\phi_{conv}(X)|0> = \int d^3p/[(2\pi)^3 2\omega_p] \qquad (5.72)$$

In "Two-Tier" quantum field theory we find the vacuum expectation value of a free field is zero *and the expectation value of the square of the field is also zero:*

$$<0|\phi(X)\phi(X)|0> = \int d^3p\, e^{-p^ip^j\Delta_{Tij}(0)/Mc^4}/[(2\pi)^3 2\omega_p] = 0 \quad (5.73)$$

since the exponential factor in the integral is $-\infty$. The exponent contains

$$\Delta_{Tij}(z) = \int d^3k\, e^{-ik.z}\,(\delta_{ij} - k_ik_j/\mathbf{k}^2)/[(2\pi)^3 2\omega_k] \quad (5.74)$$

where "T"is for "Two-Tier". Thus *vacuum fluctuations are zero in Two-Tier quantum field theory.* Correspondingly, we will see that renormalization constants are finite in the Two-Tier versions of QED, Electroweak Theory, the Standard Model and Quantum Gravity.

5.10 The Feynman Propagator

The Feynman propagator for a Two-Tier free scalar quantum field is:

$$i\Delta_F^{TT}(y_1 - y_2) = <0|T(\phi(X(y_1)),\phi(X(y_2)))|0> \quad (5.75)$$

$$\equiv <0|\phi(X(y_1))\phi(X(y_2))|0>\,\theta(y_1^{\,0} - y_2^{\,0}) +$$

$$+ \phi(X(y_2))\phi(X(y_1))|0>\,\theta(y_2^{\,0} - y_1^{\,0}) \quad (5.76)$$

Since $X^0 = y^0$ in the Coulomb gauge of the X^μ field there is no ambiguity in the choice of the relevant time variable. A straightforward calculation shows:

$$i\Delta_F^{TT}(y_1 - y_2) = i\int d^4p\, e^{-ip\cdot(y_1 - y_2)}\, R(\mathbf{p}, y_1 - y_2)/[(2\pi)^4(p^2 - m^2 + i\varepsilon)] \quad (5.77)$$

where

$$R(\mathbf{p}, y_1 - y_2) = \exp[-p^ip^j\Delta_{Tij}(y_1 - y_2)/M_c^{\,4}] \quad (5.78)$$

$$= \exp\{-p^2[A(v) + B(v)\cos^2\theta]/[4\pi^2 M_c^{\,4} z^2]\} \quad (5.79)$$

with

$$z^\mu = y_1^{\,\mu} - y_2^{\,\mu} \quad (5.80)$$
$$z = |\mathbf{z}| = |\mathbf{y_1} - \mathbf{y_2}| \quad (5.81)$$
$$p = |\mathbf{p}| \quad (5.82)$$
$$v = |z^0|/z \quad (5.83)$$

$$A(v) = (1 - v^2)^{-1} + .5v\ln[(v-1)/(v+1)] \quad (5.84)$$

$$B(v) = v^2(1 - v^2)^{-1} - 1.5v\ln[(v-1)/(v+1)] \quad (5.85)$$

$$\mathbf{p}\cdot\mathbf{z} = \text{pz}\cos\theta \qquad (5.86)$$

and $|\mathbf{p}|$ denoting the length of a spatial vector **p** while $|z^0|$ is the absolute value of z^0.

The Gaussian damping factor R(p, z) for large momentum p is the same for both the positive and negative frequency parts of the Two-Tier Feynman propagator. It is also important to note that R(p, z) does not depend on p^0 (in the Y Coulomb gauge) and thus the integration over p^0 proceeds in the usual way to produce time-ordered positive and negative frequency parts.

5.11 Large Distance Behavior of Two-Tier Theories

The large distance behavior of the Two-Tier Feynman propagator approaches the behavior of the conventional Feynman propagator since

$$R(\mathbf{p}, y_1 - y_2) \rightarrow 1 \qquad (5.87)$$

when $(y_1 - y_2)^2$ becomes much larger than M_c^{-2}. Thus the behavior of a conventional quantum field theory naturally emerges at large distance. We will see that the conventional Standard Model is the large distance limit of the Two-Tier Standard Model thus *realizing a form of Correspondence Principle for Quantum Field Theory.* Some features of the conventional Standard Model that depend specifically on the existence of divergences, such as the axial anomaly, will be different in the Two-Tier Standard Model since it is a divergence-free theory.

5.12 Short Distance Behavior of Two-Tier Theories

At short distances the Gaussian factor dominates, and radically changes, the behavior of the Feynman propagator eliminating its short distance singular behavior, and thus paving the way to finite quantum field theories. Near the light cone, $M_c^{-2} \gg -(y_1 - y_2)^2 \rightarrow 0$, we can approximate eq. 5.77 with

$$i\Delta_F^{TT}(y_1 - y_2) \approx \int d^3p\, [N(p)]^2\, R(\mathbf{p}, y_1 - y_2) \qquad (5.88)$$

since $e^{-ip\cdot(y_1 - y_2)}$ is approximately unity for small $(y_1 - y_2)$. We assume the mass of the ϕ particle is zero or is negligible at high energies so we set m = 0 to study the high energy behavior of eq. 5.88. Upon performing the integrations in eq. 5.88 for space-like $(y_1 - y_2)^2$ (and analytically continuing to the time-like regions[30,31]) we find

$$i\Delta_F^{TT}(y_1 - y_2) \approx [z^2M_c^4/(4i\sqrt{A}\sqrt{B})] \ln[(\sqrt{A} + i\sqrt{B})/(\sqrt{A} - i\sqrt{B})] \quad (5.89)$$

---

[30] See S. Blaha, "Relativistic Bound State Models with Quasi-Free Constituent Motion", Phys. Rev. **D12**, 3921 (1975) and references therein.

[31] It should be noted that A and B in eqs. 4.84-85 have the same sign for $0 \leq v < 1.1243$ thus making for easy analytic continuation across the light cone.

with A and B defined in eqs. 5.84 and 5.85. As $(y_1 - y_2)^2 \rightarrow 0$ from the space-like or time-like side of the light cone we find eq. 5.89 becomes:

$$i\Delta_F{}^{TT}(y_1 - y_2) \rightarrow \pi M_c{}^4|(y_1 - y_2)^\mu(y_1 - y_2)_\mu|/8 \qquad (5.90)$$

Eq. 5.90 has several noteworthy points:

1. The propagator is well behaved on the light cone and approaches zero smoothly from both space-like and time-like directions. In contrast, the conventional scalar Feynman propagator diverges as $[(y_1 - y_2)^\mu(y_1 - y_2)_\mu]^{-2}$. This good behavior near the light cone will be seen later for other particle propagators with the net result that the usual infinities found in conventional quantum field theory are absent in Two-Tier quantum field theories.

2. The quadratic form of the propagator in eq. 5.90 is suggestive of attempts to formulate a relativistic harmonic oscillator model of elementary particles[32] and more recent attempts to achieve quark confinement. The fact that the absolute value of the quadratic term appears in eq. 5.90 neatly avoids the common pitfall seen in fully relativistic harmonic oscillator attempts.

3. The quadratic behavior *in coordinate space* of the propagator at short distances is equivalent to a high-energy behavior of

$$p^{-6} \qquad (5.91)$$

in momentum space. Thus we get the equivalent *of a higher derivative theory* in Two-Tier quantum field theory at high energies while retaining a positive definite energy spectrum. The problems of negative metric states that have plagued conventional higher derivative quantum field theories are avoided.[33]

---

[32] H. Yukawa, H., Phys. Rev. **91**, 416 (1953); Y. S. Kim and M. E. Noz, Phys. Rev. **D8**, 3521 (1973) and references therein.

[33] S. Blaha, Phys.Rev. **D10**, 4268 (1974); S. Blaha, Phys.Rev. **D11**, 2921 (1975); S. Blaha, Nuovo Cim. **A49**, :113 (1979); S. Blaha, "Generalization of Weyl's Unified Theory to Encompass a Non-Abelian Internal Symmetry Group" SLAC-PUB-1799, Aug 1976; S. Blaha, "Quantum Gravity and Quark Confinement" Lett. Nuovo Cim. **18**, 60 (1977); Nakanishi, N., Suppl. Prog. Theo. Phys. **51**, 1 (1972); and references therein.

6. Two-Tier Perturbation Theory

6.1 Introduction

The form of quantum field theory that we have developed in chapter 4 builds on the interplay of the flatverse with our universe. It furnishes the basis of new formulations of QED, Electroweak Theory, The Standard Model and a divergence-free, unified theory of all the known interactions.

The development of these theories requires a number of topics be addressed. This chapter develops a "bare" elementary particle theory resident in the flatverse that is transformed to a "dressed" particle theory of our universe. This combination is used in perturbation theory for interactions in our universe (assumed flat in this discussion). As much as possible, we attempt to retain the features of the standard approach. The perturbation theory that we will develop will be shown to be identical to the divergence-free perturbation theory that we developed using a path integral formalism in Blaha (2005a). This theory was shown to fully satisfy unitarity in Blaha (2005a).

6.2 An Auxiliary Asymptotic Field – The Flatverse Field

The definition of the asymptotic "free" in and out states is an issue in Two-Tier quantum field theory because the "free particle field" of the theory $\phi(X(y))$ is a "dressed" particle, *ab initio*, since it is cloaked in a cloud of Y particles as discussed in Blaha (2005a).

While one could use $\phi(X(y))$ directly to define in and out asymptotic states it is more convenient initially to introduce an auxiliary asymptotic quantum field $\Phi(y)$ resident conceptually in the flatverse that will represent "bare ϕ particle" in and out states.

We will consider the case of a scalar field. We define a bare free, scalar Klein-Gordon particle field $\Phi(y)$ in the flatverse with the physical mass m of the physical $\phi(X(y))$ particle in our universe.

$$\Phi(y) = \int d^3p\, N_m(p)\, [a(p)\, e^{-ip\cdot y} + a^\dagger(p)\, e^{ip\cdot y}] \tag{6.1}$$

using the creation and annihilation operators of $\phi(X(y))$. The set of particle states of $\Phi(y)$ has the familiar Fock space form

$$|\, p_1, p_2, \ldots p_n\rangle = a^\dagger(p_1,)a^\dagger(p_2) \ldots a^\dagger(p_n)|0\rangle \tag{6.2}$$

with powers of creation operators allowed since Φ particles are bosons. The set of particle states constitutes a complete orthonormal set of states. The corresponding bra states are defined by Hermitian conjugation:

$$<p_1, p_2, \ldots p_n| = (| p_1, p_2, \ldots p_n>)^\dagger \qquad (6.3)$$

We note that the energy spectrum of these states is positive definite with the Hamiltonian

$$H_\Phi = P_\Phi^{\ 0} = \int d^3y\ \tfrac{1}{2}\,[\pi_\Phi^{\ 2} + (\partial\Phi/\partial y^i)^2 + m^2\Phi^2] \qquad (6.4a)$$

$$= \int d^3p\ (\mathbf{p}^2 + m^2)^{1/2} a^\dagger(p)a(p) \qquad (6.4b)$$

and momentum vector:

$$\mathbf{P}_\Phi = \int d^3p\ \mathbf{p}\ a^\dagger(p)a(p) \qquad (6.5)$$

We will use this set of energy-momentum eigenstates to define asymptotic "in" and "out" states in perturbation theory.

6.3 Transformation Between Φ(y) and ϕ(X(y))

Since the y coordinates are the coordinates of the flatverse it is reasonable to view the Φ(y) fields as fields in the flatverse. These fields undergo a transformation to their equivalent in our universe via a transformation between the in and out Φ(y) fields, and the in and out $\phi(X(y))$ fields in perturbation theory.

We define a transformation $W_a(y)$ that transforms in and out Φ(y) fields to in and out $\phi(X(y))$ fields respectively:

$$\phi_a(X(y)) = :W_a(y)\Phi_a(y)W_a^{\ -1}(y): \qquad (6.6)$$

where the label a = "in" or a = "out", where : … : signifies normal ordering, and where

$$\Phi_{in}(y) = \int d^3p\ N_m(p)\ [a_{in}(p)\ e^{-ip\cdot y} + a_{in}^{\ \dagger}(p)\ e^{ip\cdot y}] \qquad (6.7)$$

$$\Phi_{out}(y) = \int d^3p\ N_m(p)\ [a_{out}(p)\ e^{-ip\cdot y} + a_{out}^{\ \dagger}(p)\ e^{ip\cdot y}] \qquad (6.8)$$

$$\phi_{in}(X) = \int d^3p\ N_m(p)\ [a_{in}(p)\ :e^{-ip\cdot(y + iY/M_c^2)}: + a_{in}^{\ \dagger}(p)\ :e^{ip\cdot(y + iY/M_c^2)}:] \qquad (6.9)$$

$$\phi_{out}(X) = \int d^3p\ N_m(p)\ [a_{out}(p)\ :e^{-ip\cdot(y + iY/M_c^2)}: + a_{out}^{\ \dagger}(p)\ :e^{ip\cdot(y + iY/M_c^2)}:] \qquad (6.10)$$

Note that the transformation eq. 6.6 includes normal ordering. While this transformation may seem strange it is no stranger than the time reversal operator, in which the complex conjugate of all c-number terms is taken in addition to applying a unitary transformation.

The $W_a(y)$ operator implements the transition of fields from the flatverse to our universe. It corresponds to the relation between flatverse coordinates and coordinates in our universe. $W_a(y)$ cloaks flatverse fields with Y quanta.

In the Coulomb gauge of Y it is easy to show that

$$W_a(y) = \exp(-\mathbf{Y}(y)\cdot\mathbf{P}_{\Phi a}/M_c^2) \qquad (6.11)$$

and

$$W_a^{-1}(y) = \exp(\mathbf{Y}(y)\cdot\mathbf{P}_{\Phi a}/M_c^2) \qquad (6.12)$$

where the label a = "in" or a = "out", where the inner products in the exponentials are the usual spatial vector inner product, and where

$$\mathbf{P}_{\Phi a} = -\int d^3y\, \partial\Phi_a(y)/\partial y^0\, \nabla\Phi_a(y) = \int d^3p\, \mathbf{p}\, a_a^\dagger(p)a_a(p) \qquad (6.12a)$$

is a spatial vector (the Φ spatial momentum operator) that is written solely in terms of $\Phi_a(y)$'s creation and annihilation operators.

In addition to performing the transformation in eq. 6.6 $W_a(y)$ also performs a "translation" in Y^μ:

$$W_a(y)Y^i(y')W_a^{-1}(y) = Y^i(y') + i\Delta^{trij}(y' - y)P_{\Phi a}^j/M_c^2 \qquad (6.13a)$$

where

$$i\Delta^{trij}(y' - y) = \int d^3k\, (e^{-ik\cdot(y' - y)} - e^{ik\cdot(y' - y)})(\delta_{jk} - k_jk_k/\mathbf{k}^2)/[(2\pi)^3 2\omega_k] \qquad (6.13b)$$

We note that $W_a(y)$ is not a unitary operator but it is pseudo-unitary:

$$W_a^{-1}(y) = VW_a^\dagger(y)V^{-1} = VW_a(y)V^{-1} \qquad (6.14)$$

where

$$V = \exp(-i\pi \sum_{\lambda=1}^{2} \int d^3k\, a^\dagger(k, \lambda)a(k, \lambda)) \qquad (6.15)$$

is a unitary operator with the property

$$VY^j(y)V^{-1} = -Y^j(y) \qquad (6.16)$$

for j = 1,2,3. We note

$$V^\dagger = V^{-1} = V \qquad (6.17)$$

and thus

$$V^2 = I \qquad (6.18)$$

V will be shown to be a metric operator in the following discussion.[34] We note that the Y "particle" (Hermitian) number operator appears in eq. 6.9 in the expression for V:

$$N_Y = \sum_{\lambda=1}^{2} \int d^3k\, a^\dagger(k, \lambda)a(k, \lambda) \qquad (6.19)$$

[34] P. A. M. Dirac, Proc. R. Soc. London A **180**, 1 (1942); T. D. Lee and G. C. Wick, Nucl. Phys. **B9**, 209 (1969); C. M. Bender, D. C. Brody and H. F. Jones, "Complex Extension of Quantum Mechanics" Phys. Rev. Letters **89**, 270401-1 (2002) and references therein.

Thus states with an even number of Y "particles" have a V eigenvalue of one, and states with an odd number of Y "particles" have a V eigenvalue of minus one.

6.4 Model Lagrangian with ϕ^4 Interaction

We will develop our perturbation theory using a scalar Lagrangian model with a ϕ^4 interaction term:

$$\mathscr{L}_S = J\mathscr{L}_F + \mathscr{L}_C \tag{6.20}$$

with

$$\mathscr{L}_F = ½ [(\partial\phi/\partial X^\nu)^2 - m^2\phi^2] + \mathscr{L}_{Fint} \tag{6.21}$$

and

$$\mathscr{L}_C = -¼ F_Y^{\mu\nu} F_{Y\mu\nu} \tag{6.22}$$

with

$$F_{Y\mu\nu} = \partial Y_\mu/\partial y^\nu - \partial Y_\nu/\partial y^\mu \tag{6.23}$$

and

$$\mathscr{L}_{Fint} = 1/4!\ \chi_0\ \phi(X(y))^4 + ½ (m^2 - m_0^2)\phi^2 \tag{6.24}$$

where J is the Jacobian, χ_0 is the bare coupling constant, and m_0 is the bare mass.

The conserved momentum operator is:

$$P_{F\beta} = \int d^3X\ \mathscr{T}_{F0\beta} \tag{6.25}$$

where

$$\mathscr{T}_{F\mu\nu} = - g_{\mu\nu} \mathscr{L}_F + \partial\mathscr{L}_F / \partial(\partial\phi/\partial X_\mu)\ \partial\phi/\partial X^\nu \tag{6.26}$$

is the conserved ϕ field energy-momentum tensor:

$$\partial P_{F\beta}/\partial X^0 = 0 \tag{6.27}$$

The Hamiltonian density is

$$\mathscr{H}_F = \mathscr{T}_{F0\beta} = \mathscr{H}_{F0} + \mathscr{H}_{Fint} \tag{6.28}$$

with

$$\mathscr{H}_{F0} = ½ [\pi_\phi^2 + (\partial\phi/\partial X^i)^2 + m^2\phi^2] \tag{6.29}$$

$$\mathscr{H}_{Fint} = - 1/4!\ \chi_0\ \phi(X(y))^4 + ½ (m^2 - m_0^2)\phi(X(y))^2 \tag{6.30}$$

6.5 In-states and Out-States

In this section we will develop properties of in-fields and out-fields. We will use a two-step process to set up the perturbation theory for the S matrix. The first step defines a free field in-field and out-field formalism in flatverse coordinates y. The second step is the transformation of the flatverse in-fields and out-fields to free in-fields and out-fields in our universe where the coordinates are X(y). The process can be schematized as:

$$\Phi_{in}(y) \Rightarrow \phi_{in}(X(y)) \Rightarrow \phi(X(y)) \Rightarrow \phi_{out}(X(y)) \Rightarrow \Phi_{out}(y) \quad (6.31)$$

In-states are constructed using the flatverse field Φ_{in} which are then effectively transformed into $\phi_{in}(X(y))$ expressions in order to make contact with our Lagrangian formalism. Then $\phi_{in}(X(y))$ is related to the interacting field $\phi(X(y))$ as a limit ($y^0 \to -\infty$). Similarly out-states are constructed using the flatverse field Φ_{out} which are then expressed in terms of $\phi_{out}(X(y))$. Then $\phi_{out}(X(y))$ is related to the interacting field $\phi(X(y))$ using the LSZ limiting process ($y^0 \to +\infty$).

Since much of the development differs only trivially from the standard treatment in textbooks we will simply "list" relevant equations and let the reader pursue them further in quantum field theory textbooks as needed.

6.6 Our Universe ϕ In-Field

In order to define a perturbation theory for particle scattering we will next specify features of the in-field $\phi_{in}(X(y))$ and in-field states – the field and states representing physical particles as $X^0 = y^0 \to -\infty$.[35]

A. The in-field $\phi_{in}(X(y))$ satisfies the Klein-Gordon equation in the X variable:

$$(\Box_X + m^2)\, \phi_{in}(X) = 0 \quad (6.32)$$

where

$$\Box_X = (\partial/\partial X^\nu)(\partial/\partial X_\nu)$$

B. Under coordinate displacements and Lorentz transformations $\Phi_{in}(y)$, $\phi_{in}(X(y))$, and $\phi(X(y))$ transform in the same way in their respective spaces:

$$[P^\mu, \Phi_{in}(y)] = -\, i\partial\Phi_{in}/\partial y_\mu \quad (6.33a)$$
$$[P^\mu, \phi_{in}(X)] = -\, i\partial\phi_{in}/\partial y_\mu \quad (6.33b)$$
$$[P^\mu, \phi(X)] = -\, i\partial\phi/\partial y_\mu \quad (6.34)$$

with the energy-momentum vector P^μ.

---

[35] There is an implicit Wick rotation of y_0 to make it explicitly a time coordinate. Since y_0 is complex there are no issues.

C. We can relate the asymptotic in-field $\phi_{in}(X(y))$ to the interacting field $\phi(X(y))$ using the equation of motion of $\phi(X(y))$

$$(\Box_X + m^2)\,\phi(X) = j(X) \tag{6.35}$$

where j(X) embodies the interaction. Using the physical mass m we find

$$(\Box_X + m^2)\,\phi(X) = j(X) + (m^2 - m_0{}^2)\phi(X) = j_{tot}(X) \tag{6.36}$$

If the current is taken to be the source of the scattered waves we may write

$$\sqrt{Z}\,\phi_{in}(X(y)) = \phi(X(y)) - \int d^4X(y')\,\Delta_{ret}(y - y')\,j_{tot}(X(y')) \tag{6.37}$$
$$= \phi(X(y)) - \int d^4y'\,J\,\Delta_{ret}(y - y')\,j_{tot}(X(y')) \tag{6.38}$$

where Z is a wave function renormalization constant, J is the Jacobian, and Δ_{ret} is a retarded Green's function.

D. We can define $\Phi_{in}(y)$ in-field states with expressions like

$$|\,p_1, p_2, \ldots\, p_n\ in\rangle = a_{in}{}^\dagger(p_1)a_{in}{}^\dagger(p_2)\,\ldots\, a_{in}{}^\dagger(p_n)|0\rangle \tag{6.39}$$

with powers of creation operators allowed since Φ_{in} is a boson field. The set of all particle states constitutes a complete orthonormal set of states. The corresponding bra states are defined by Hermitian conjugation:

$$\langle p_1, p_2, \ldots\, p_n\ in| = (|\,p_1, p_2, \ldots\, p_n\ in\rangle)^\dagger \tag{6.40}$$

6.7 Our Universe ϕ Out-Field

In order to define a perturbation theory for particle scattering we begin by listing aspects of the out-field $\phi_{out}(X(y))$ and out-field states – the field and states representing physical particles as $X^0 = y^0 \to -\infty$.

A. The out-field $\phi_{out}(X(y))$ satisfies the Klein-Gordon equation in the X variable:

$$(\Box_X + m^2)\,\phi_{out}(X) = 0 \tag{6.41}$$

where

$$\Box_X = (\partial/\partial X^\nu)(\partial/\partial X_\nu)$$

B. Under coordinate displacements and Lorentz transformations $\Phi_{out}(y)$, $\phi_{out}(X(y))$, and $\phi(X(y))$ transform in the same way:

$$[P^\mu, \Phi_{out}(y)] = -\,i\partial\Phi_{out}/\partial y_\mu \tag{6.42a}$$

$$[P^\mu, \phi_{out}(X)] = -i\partial\phi_{out}/\partial y_\mu \qquad (6.42b)$$
$$[P^\mu, \phi(X)] = -i\partial\phi/\partial y_\mu \qquad (6.43)$$

with the energy-momentum vector P^μ.

C. We can relate the asymptotic out-field $\phi_{out}(X(y))$ to the interacting field $\phi(X(y))$ using the equation of motion of $\phi(X(y))$ specified by eq. 6.36:

$$\sqrt{Z}\,\phi_{out}(X(y)) = \phi(X(y)) - \int d^4X(y')\,\Delta_{adv}(y - y')\,j_{tot}(X(y')) \qquad (6.44)$$
$$= \phi(X(y)) - \int d^4y'\,J\,\Delta_{adv}(y - y')\,j_{tot}(X(y')) \qquad (6.45)$$

where Z is a wave function renormalization constant, J is the Jacobian, and Δ_{adv} is an advanced Green's function.

D. We can define Φ_{out} out-field states with expressions like

$$|\,p_1, p_2, \ldots\, p_n\ out> = a_{out}{}^\dagger(p_1,)a\Phi_{out}{}^\dagger(p_2)\ \ldots\ a\Phi_{out}{}^\dagger(p_n)|0> \qquad (6.46)$$

with powers of creation operators allowed since Φ_{out} is a boson field. The set of all particle states constitutes a complete orthonormal set of states. The corresponding bra states are defined by Hermitian conjugation:

$$<p_1, p_2, \ldots\, p_n\ out| = (|\,p_1, p_2, \ldots\, p_n\ out>)^\dagger \qquad (6.47)$$

6.8 The Y Field

The Y field in the present model Lagrangian (eq. 6.20) is a free field and thus:

$$Y_{in}(y) = Y_{out}(y) = Y(y) \qquad (6.48)$$

The states of the Y field have two general forms: 1) States in a Fock space consisting of particle states that are eigenstates of the Y particle number operator (eq. 6.19); and 2) Coherent states in a non-Fock space of generalized coherent states in an infinite tensor product space.[36]

The coherent ket states that arise in Two-Tier quantum field theory have the general form:

$$|y,\ p> = e^{-\mathbf{p}\cdot\mathbf{Y}^-(y)/M_c{}^2}|0>$$

as can be seen from an examination of $\phi_{in}(X(y))$. The corresponding bra state is:

---

[36] See Kibble and other references on coherent states.

$$<y, p| = (V| y, p>)^{\dagger} = <0|e^{+\mathbf{p}\cdot\mathbf{Y}^{+}(y)/M_c^{2}} \qquad (6.49)$$

with V, the metric operator, reversing the sign of Y in the exponential. The inner product of coherent states is:

$$<y_1 , p_1| y_2, p_2> = \exp[-p_1^{\,i} p_2^{\,j} \Delta_{Tij}(y_1 - y_2)/M_c^{4}] \qquad (6.50)$$

showing the set of coherent states is not orthonormal and, in fact, is overcomplete. Comparing eq. 6.50 to eq. 5.78 gives

$$<y_1 , p| y_2, p> = R(p, y_1 - y_2) \qquad (6.50a)$$

The completeness of the set of states for each time y^0 can be verified by examining the projection operator:

$$\mathscr{R}_Y(y^0) = \because \exp[-i \int d^3y\, Y^{-}{}_i(y)|0><0|\pi^{+j}(y)] \because \qquad (6.51)$$

where

$$\pi^{+j}(y) = -\partial Y^{+j}(y)/\partial y^0 \qquad (6.52)$$

and where ∵ represents an extended normal ordering operator:

$$\because \ldots \because$$

which is defined as placing creation operators to the left, projection operators in the center, and annihilation operators to the right. Thus eq 6.51 can be written

$$\mathscr{R}_Y = \Sigma_n (-i/n!)^n \int d^3y_1 \ldots \int d^3y_n Y^{-j_1}(y_1) Y^{-j_2}(y_2) \ldots Y^{-j_n}(y_n)|0><0|\pi^{+}{}_{j_1}(y_1)\pi^{+}{}_{j_2}(y_2)\ldots\pi^{+}{}_{j_n}(y_n) \qquad (6.53)$$

where we have used the fact that |0><0| is a projection operator, and reduced |0><0| |0><0| … |0><0| to |0><0| in eq. 6.53. The vacuum state is the product of the Y and ϕ vacuum states:

$$|0> = |0_Y>|0_\phi> \qquad (6.53a)$$

We note

$$\mathscr{R}_Y (y^0)|\mathbf{y},y^0\ p> = |\mathbf{y},y^0\ p> \qquad (6.54)$$

and $\int d^3y_2\, p^i\, \Delta^{tr}{}_{ij}(y_1 - y_2) Y^{+j}(y_2) = \mathbf{p}\cdot\mathbf{Y}^{+}(y_1)$. Also

$$\mathscr{R}_Y(y^0)|n> = |n> \qquad (6.55)$$

where |n> is any Y particle Fock state of finite particle number. In view of eqs. 6.54 and 6.55, we see that $\mathscr{R}_Y$ is the identity operator in the Fock space and in the space of generalized coherent Y field states. Thus the set of Y coherent states forms an over complete set of states. We will define the S matrix for any combination of Φ Fock space states and coherent Y states. The $\mathscr{R}_Y$ operator can be generalized to include Φ Fock space states:

$$\mathscr{R}_{\Phi Y}(y^0) = \because \exp[-i \int d^3y Y^-_j(y) R_\Phi \pi^{+j}(y)] \because \quad (6.56)$$

with

$$\mathscr{R}_\Phi = \Sigma_n |n><n| \quad (6.57)$$

being a sum over all Φ Fock space states with vacuum state given by eq. 6.53a. Since $\mathscr{R}_\Phi$ is a projection:

$$[R_\Phi]^N = R_\Phi$$

for any power N, we find:

$$\mathscr{R}_{\Phi Y}(y^0) = \Sigma_n (-i)^n \int d^3y_1 \ldots \int d^3y_n Y^{-j_1}(y_1) Y^{-j_2}(y_2) \ldots Y^{-j_n}(y_n) \mathscr{R}_\Phi \pi^+_{j_1}(y_1) \pi^+_{j_2}(y_2) \ldots \pi^+_{j_n}(y_n) \quad (6.58)$$

As a result we have

$$\mathscr{R}_{\Phi Y}(y^0)|y, p; n_\Phi> = |y, p; n_\Phi> \quad (6.59)$$

for any combination of Y coherent states and Φ Fock space states n_Φ. Also

$$\mathscr{R}_{\Phi Y}(y^0)|n_\Phi> = |n_\Phi> \quad (6.60)$$

Thus $\mathscr{R}_{\Phi Y}$ is the identity operator on this space – the (over) complete space of in and out states which we will use to define the S matrix of the scalar field theory specified by the Lagrangian eq. 6.20.

6.9 S Matrix

Following the standard definition of the S matrix we have:

$$S_{\alpha\beta} = <\alpha \text{ out}|\beta \text{ in}> \quad (6.61)$$
$$= <\alpha \text{ in}|S|\beta \text{ in}> \quad (6.62)$$

$$|0> = |0 \text{ in}> = |0 \text{ out}> = S|0 \text{ in}> \quad (6.63)$$

$$\Phi_{in}(y) = S\Phi_{out}(y)S^{-1} \quad (6.64)$$

and the other standard properties of the S matrix with the sole exception being the form of the unitarity relation (discussed later).

6.10 LSZ Reduction for Scalar Fields

In this section we determine the reduction formula for the S matrix for scalar ϕ fields. Consider the S matrix element corresponding to an in state of particles β plus one ϕ particle of momentum p, and an out state α:

$$S_{\alpha\beta p} = \langle\alpha\ \text{out}|\beta p\ \text{in}\rangle \tag{6.65}$$

After standard manipulations we have:

$$S_{\alpha\beta p} = \langle\alpha - p\ \text{out}|\beta\ \text{in}\rangle - i\langle\alpha\ \text{out}|\int d^3y\ f_p(y)\ \overleftrightarrow{\partial_0}\ [\Phi_{in}(y) - \Phi_{out}(y)]\ |\beta\ \text{in}\rangle \tag{6.66}$$

where $\langle\alpha - p\ \text{out}|$ is an out state with a particle of momentum p removed (if present) and where

$$f(y^0)\ \overleftrightarrow{\partial_0}\ g(y^0) = f(y^0)\ \partial g(y^0)/\partial y^0 - \partial f(y^0)/\partial y^0\ g(y^0) \tag{6.67}$$

and

$$f_p(y) = N_m(p)e^{-ip\cdot y} \tag{6.68}$$

with $N_m(p)$ a normalization constant.
We now express

$$S_{\alpha\beta p} = S_{\alpha-p\beta} - i\langle\alpha\ \text{out}|\ \int d^3y\ f_p(y)\ \overleftrightarrow{\partial_0}\ W^{-1}[\phi_{in}(X(y)) - \phi_{out}(X(y))]W|\beta\ \text{in}\rangle \tag{6.69}$$

using $W(y) = W_{in}(y)$ with

$$\Phi_a(y) = W_a^{-1}(y)\phi_a(X(y))W_a(y) \tag{6.70}$$

where the label a = "in" or a = "out", and where

$$W_a(y) = \exp(-\mathbf{Y}(y)\cdot\mathbf{P}_{\Phi a}/M_c^2) \tag{6.71}$$

and

$$W_a^{-1}(y) = \exp(\mathbf{Y}(y)\cdot\mathbf{P}_{\Phi a}/M_c^2) \tag{6.72}$$

in the Coulomb gauge of Y with $\mathbf{P}_{\Phi a}$ the momentum spatial vector defined by eq. 6.12a.

We note that the interacting $\phi(X(y))$ approaches the in and out fields $\phi_{in}(X(y))$ and $\phi_{out}(X(y))$ in the limit that $y^0 \to -\infty$ and $y^0 \to +\infty$ respectively in the sense of Lehmann, Symanzik and Zimmermann[37] which we *symbolize* as:

$$\phi(X(y)) \to \surd Z\, \phi_{in}(X(y)) \quad \text{as} \quad y^0 \to -\infty \quad (6.73)$$

$$\phi(X(y)) \to \surd Z\, \phi_{out}(X(y)) \quad \text{as} \quad y^0 \to +\infty \quad (6.74)$$

with $\surd Z$ defined in eqs. 6.37 and 6.44. Thus we can rewrite eq. 6.69 as

$$S_{\alpha\beta p} = S_{\alpha-p\beta} + iZ^{-1/2} (\lim_{y^0 \to +\infty} - \lim_{y^0 \to -\infty}) <\alpha \text{ out}| \int d^3y\, f_p(y) \overleftrightarrow{\partial_0} W^{-1}\phi(X(y))W|\beta \text{ in}> \quad (6.75)$$

which, after standard manipulations, becomes

$$S_{\alpha\beta p} = S_{\alpha-p\beta} + iZ^{-1/2} \int d^4y\, f_p(y)(\Box_y + m^2)<\alpha \text{ out}|\, W(y)^{-1}\phi(X(y))W(y)|\beta \text{ in}> \quad (6.76)$$

Eq. 6.76 is similar to the usual LSZ reduction formula except for the appearance of the W(y) operator and its inverse. We note that $W(y) = W_{in}(y)$ still because $\mathbf{P}_{\Phi in}$ is independent of y^0.

Similarly an out ϕ particle can be reduced from an S matrix part. For example,

$$\begin{aligned}<\alpha \text{ out}|W^{-1}(y)\phi(X(y))W(y)|\beta \text{ in}> &= <\alpha - p' \text{ out}|W^{-1}(y)\phi(X(y))W(y)|\beta - p' \text{ in}> \\ &- i<\alpha - p' \text{ out}| \int d^3y' [W^{-1}(y')\phi_{in}(X(y'))W(y')W^{-1}(y)\phi(X(y))W(y) - \\ &- W^{-1}(y)\phi(X(y))W(y)W^{-1}(y')\phi_{out}(X(y'))W(y')]|\beta \text{ in}> \overleftrightarrow{\partial_0} f_{p'}^*(y') \end{aligned} \quad (6.77)$$

which becomes

$$\begin{aligned}<\alpha \text{ out}|W^{-1}(y)\phi(X(y))W(y)|\beta \text{ in}> &= <\alpha - p' \text{ out}|\varphi(y)|\beta - p' \text{ in}> + \\ &+ iZ^{-1/2} \int d^4y' <\alpha - p' \text{ out}|T(\varphi(y')\varphi(y))|\beta \text{ in}> (\overleftarrow{\Box}_{y'} + m^2) f_{p'}^*(y') \end{aligned} \quad (6.78)$$

where the time ordered product is defined with respect to ordering with y^0 and where

$$\varphi(y) = W^{-1}(y)\phi(X(y))W(y) \quad (6.79)$$

These results directly generalize to multi-particle in and out states:

---

[37] H. Lehmann, K. Symanzik and W. Zimmermann, Nuov. Cim., **1**, 1425 (1955); W. Zimmermann, Nuov. Cim., **10**, 567 (1958); O. W. Greenberg, Doctoral Dissertation, Princeton University 1956.

$$<p_1, p_2, \ldots p_n \text{ out}| q_1, q_2, \ldots q_m \text{ in}> =$$
$$= \ldots <0|T(\varphi(y'_1) \ldots \varphi(y'_n)\varphi(y_1) \ldots \varphi(y_m))|0> \ldots \quad (6.80)$$

thus reducing the development of the perturbation theory of the S matrix to the evaluation of time ordered products such as

$$<0|T(\varphi(y_1) \ldots \varphi(y_n))|0> \quad (6.81)$$

6.11 The U Matrix

The U matrix for a Two-Tier theory is developed in a way similar to conventional field theory starting from the defining relations:

$$\phi(X(y)) = U^{-1}\phi_{in}(X(y))U \quad (6.82)$$
$$\pi_\phi(X(y)) = U^{-1}\pi_{\phi in}(X(y))U \quad (6.83)$$

From eq. 6.29 we define the free field Hamiltonian

$$H_{F0in}(\phi_{in}, \pi_{\phi in}) = \int d^3X\, \mathscr{H}_{F0}(\phi_{in}, \pi_{\phi in}) \quad (6.84)$$

Noting $X^0 = y^0$ in the Y Coulomb gauge we find

$$\partial\phi_{in}/\partial y^0 = i[H_{F0in}, \phi_{in}(X)] \quad (6.85)$$
$$\partial\pi_{\phi in}/\partial y^0 = i[H_{F0in}, \pi_{\phi in}(X)] \quad (6.86)$$

For the entire Hamiltonian (eq. 6.28) we have

$$\partial\phi/\partial y^0 = i[H_F, \phi(X)] \quad (6.87)$$
$$\partial\pi_\phi/\partial y^0 = i[H_F, \pi_\phi(X)] \quad (6.88)$$

with

$$H_F(\phi, \pi_\phi) = :\int d^3X\, \mathscr{H}_F(\phi, \pi_\phi): \quad (6.89)$$

(Note the *entire* interaction term is normal ordered since d^3X is a q-number. Combining the above equations in the standard way yields a familiar differential equation for the U matrix:

$$i\partial U(y^0)/\partial y^0 = (H_{Fint} + E_0(t))U(y^0) \quad (6.90)$$

where $E_0(t)$ is a c-number function of y^0 that we can set equal to 0 (as it would be cancelled later in any case), and where

$$H_{Fint}(\phi_{in}, \pi_{\phi in}) = :\int d^3X\, \mathscr{H}_{Fint}(\phi_{in}, \pi_{\phi in}): \quad (6.91)$$

with $\mathscr{H}_{Fint}$ given by eq. 6.30. Solving for U gives the familiar time ordered exponential:

$$U(y^0) = T(\exp[-i \int_{-\infty}^{t} dy^0\, H_{Fint}]) \qquad (6.92)$$

which is symbolic notation for:

$$U(y^0) = 1 + \sum_{n=1}^{\infty} (-i)^{-n} (n!)^{-1} \int_{-\infty}^{y^0} dy_1{}^0 \ldots \int_{-\infty}^{y^0} dy_n{}^0\, T(H_{Fint}(y_1{}^0) \ldots H_{Fint}(y_n{}^0)) \qquad (6.93)$$

We note for later use that the hermiticity of H_{Fint} is not used in the derivation of eq. 6.93. Thus eq. 6.93 would still hold if H_{Fint} were not Hermitian.

6.12 Reduction of Time Ordered φ Products

In the previous chapter we reduced the calculation of the S matrix to the evaluation of time ordered products of the form

$$\tau(y_1, \ldots, y_n) = \langle 0|T(\varphi(y_1) \ldots \varphi(y_n))|0\rangle \qquad (6.94)$$

where $\varphi(y)$ is specified by eq. 6.79. Expanding the terms within eq. 6.94 using eq. 6.79 we find

$$\varphi(y_1) \ldots \varphi(y_n) = W^{-1}(y_1)\phi(X(y_1))W(y_1)W^{-1}(y_2)\phi(X(y_2))W(y_2) \ldots W^{-1}(y_n)\phi(X(y_n))W(y_n) \qquad (6.95)$$

which can be re-expressed as

$$W^{-1}(y_1)U^{-1}(y_1{}^0)\phi_{in}(X(y_1))U(y_1{}^0)W(y_1)W^{-1}(y_2)\cdot U^{-1}(y_2{}^0)\phi_{in}(X(y_2))U(y_2{}^0)W(y_2) \ldots \qquad (6.96)$$

denoting $W_{in}(y)$ as W(y). Defining

$$\mathscr{U}(y_1, y_2) = U(y_1{}^0)W(y_1)W^{-1}(y_2)U^{-1}(y_2{}^0) \qquad (6.97)$$

we see eq. 6.14 can be rewritten as

$$W^{-1}(y_1)U^{-1}(y_1{}^0)\phi_{in}(X(y_1))\mathscr{U}(y_1, y_2)\phi_{in}(X(y_2))\, \mathscr{U}(y_2, y_3)\phi_{in}(X(y_3)) \ldots$$
$$\ldots \phi_{in}(X(y_n))U(y_n{}^0)W(y_n) \qquad (6.98)$$

From eqs. 6.71 and 6.72

$$\mathscr{U}(y_1, y_2) = U(y_1{}^0)\exp((\mathbf{Y}(y_2) - \mathbf{Y}(y_1))\cdot\mathbf{P}_{\Phi a}/M_c{}^2)U^{-1}(y_2{}^0) \qquad (6.99)$$

bringing the flatverse expression into the equivalent expression in our universe. Defining

$$W(y_1, y_2) = \exp((\mathbf{Y}(y_2) - \mathbf{Y}(y_1))\cdot\mathbf{P}_{\Phi a}/M_c^2) \tag{6.100}$$

and looking ahead to the Wick expansion of the time ordered product of eq. 6.93 we note that the only time ordered products involving $W(y_1, y_2)$ that would appear in the expansion are

$$<0|T(\phi_{in}(X(y))W(y_1, y_2))|0> = 0 \tag{6.101}$$

$$<0|T(Y(y)W(y_1, y_2))|0> = 0 \tag{6.102}$$

$$<0|T(\partial Y(y)/\partial y^\mu\, W(y_1, y_2))|0> = 0 \tag{6.103}$$

$$<0|T(\partial Y(y)/\partial y^\mu\, \phi_{in}(X(y)))|0> = 0 \tag{6.104}$$

$$<0|T(W(y_1, y_2)W(y_3, y_4))|0> = 1 \tag{6.105}$$

due to the factor of $\mathbf{P}_{\Phi a}$ that appears in $W(y_1, y_2)$. Also

$$<0|T(\phi_{in}(X(y))Y(y_1))|0> = 0 \tag{6.106}$$

due to the $a_{in}(p)$ and $a_{in}^\dagger(p)$ factors appearing in $\phi_{in}(X(y))$.

Thus the $W(y_1, y_2)$ factor in eq. 6.99 may be set to the value one with the result

$$\mathscr{U}(y_1, y_2) \equiv U(y_1^0)U^{-1}(y_2^0) = U(y_1^0, y_2^0) \tag{6.107}$$

where $U(y_1^0, y_2^0)$ is the conventionally defined U matrix satisfying

$$i\partial U(y_1^0, y_2^0)/\partial y_1^0 = iH_{Fint}\, U(y_1^0, y_2^0) \tag{6.108}$$

with the boundary condition

$$U(y^0, y^0) = 1 \tag{6.109}$$

This result would still be true if the $W(y_1, y_2)$ exponentials were expanded in their "power series" form.

Then, paralleling the standard approach we find an expression for the U matrix:

$$U(y_1^0, y_2^0) = T(\exp[-i \int_{y_2^0}^{y_1^0} dy'^0 :d^3X(y')\, \mathscr{H}_{Fint}(\phi_{in}(X(y')), \pi_{\phi in}(X(y'))):]) \tag{6.110}$$

The $U(y_1^0, y_2^0)$ matrix satisfies the conventional multiplication rule:

$$U(y_1^0, y_3^0) = U(y_1^0, y_2^0)U(y_2^0, y_3^0) \tag{6.111}$$

The inverse of $U(y_1, y_2)$ is

$$U^{-1}(y_1^0, y_2^0) = U(y_2^0, y_1^0) \tag{6.112}$$

We now return to eq. 6.98, which can now be written in the form:

$$U^{-1}(y^0)U(y^0, y_1^0)\phi_{in}(X(y_1))U(y_1^0, y_2^0)\phi_{in}(X(y_2))U(y_2^0, y_3^0) \ldots \\ \ldots\phi_{in}(X(y_n))U(y_n^0, -y^0)U(-y^0) \tag{6.113}$$

where y^0 is a reference time that is later than all other times, and $-y^0$ is earlier than all the other times, in the time-ordered product. As a result the time-ordered product in eq. 6.80 can be expressed in a symbolic notation as:

$$<0|U^{-1}(y^0)T(\phi_{in}(X(y_1))\phi_{in}(X(y_2)) \ldots \phi_{in}(X(y_n))U(y^0, -y^0))U(-y^0)|0> \tag{6.114}$$

The analysis of eq. 6.114 as $y^0 \rightarrow \infty$ follows the standard path, which begins by noting

$$U(-y)|0> = \lambda_-|0> \qquad \text{when } y^0 \rightarrow \infty \tag{6.115}$$

$$U(y)|0> = \lambda_+|0> \qquad \text{when } y^0 \rightarrow \infty \tag{6.116}$$

following a standard textbook proof, which, in turn, leads to:

$$\lambda_-\lambda_+^* = <0|T(\exp[+i\int_{-\infty}^{\infty} dy'^0{:}d^3X(y')\mathscr{H}_{Fint}(\phi_{in}(X(y')),\pi_{\phi in}(X(y'))){:}])|0> \tag{6.117}$$

$$= [<0|T(\exp[-i\int_{-\infty}^{\infty} dy'^0 d^3X(y')\mathscr{H}_{Fint}(\phi_{in}(X(y')),\pi_{\phi in}(X(y')))])|0>]^{-1} \tag{6.118}$$

Thus the time ordered product above, which appears in the evaluation of the S matrix element in eq. 6.80, can be symbolically written as:

$$\tau(y_1, \ldots, y_n) = \frac{<0|T(\phi_{in}(X(y_1)) \ldots \phi_{in}(X(y_n))U(\infty, -\infty))|0>}{<0|T(\exp[-i\int dy'^0{:}d^3X(y')\mathscr{H}_{Fint}(\phi_{in}(X(y')),\pi_{\phi in}(X(y'))){:}])|0>} \tag{6.119}$$

in the limit $y^0 \rightarrow \infty$.

6.13 The $\int d^3X$ Integration

The integration over the X space coordinates presents the difficulty of a functional integration of a q-number that needs to be properly defined. Since

$$X^\mu(y) = y^\mu + i\, Y^\mu(y)/M_c^{\,2} \tag{6.120}$$

by definition and since, in the Y Coulomb gauge we have $X^0(y) = y^0$ due to $Y^0 = 0$, the classical Jacobian for the transformation from y to X coordinates is the absolute value:

$$J = \left| \varepsilon^{ijk}\left(\delta^{1i} + \frac{i}{M_C^{\,2}}\frac{\partial Y^1}{\partial y^i}\right)\left(\delta^{2j} + \frac{i}{M_C^{\,2}}\frac{\partial Y^2}{\partial y^j}\right)\left(\delta^{3k} + \frac{i}{M_C^{\,2}}\frac{\partial Y^3}{\partial y^k}\right)\right| \tag{6.121}$$

The Jacobian appears in a change of integration variables in the Y Coulomb gauge.:

$$\int d^3X = \int d^3y\, J \tag{6.122}$$

and

$$\int d^4X = \int d^4y\, J \tag{6.123}$$

A change of variables for c-number coordinate transformations is well known. The situation changes when one set of coordinates are in fact q-numbers. The second quantization of the Y field requires the definition of J to be clarified since the product of fields at the same position is normally undefined. The normal ordering of the interaction Hamiltonian term resolves the issue. Therefore eq. 6.123 must be considered as inserted within a normal ordered expression.

While normal ordering eliminates the infinities that would otherwise be present, J still presents a problem because it is still effectively part of the interaction term. This situation appears to be unsatisfactory in the present, scalar quantum field theory in which Y is not intended to play a direct dynamical role but rather a passive role as a coordinate. The normal ϕ field is supposed to be the only in, out, and interacting field.

The problem of J is resolved by eqs. 6.103 and 6.104, which reduces the effect of the derivative terms to zero in the Wick expansion of the time ordered product if no Y quanta appear in or out S matrix states. Thus

$$J \equiv 1 \tag{6.124}$$

As a result the time ordered product (eq. 6.119) becomes:

$$\tau(y_1,\ldots, y_n) = \frac{<0|T(\phi_{in}(X(y_1))\ldots\phi_{in}(X(y_n))\exp[-i\int d^4y\,\mathscr{H}_{Fint}(\phi_{in}(X(y')))])|0>}{<0|T(\exp[-i\int d^4y'\,\mathscr{H}_{Fint}(\phi_{in}(X(y')))])|0>} \tag{6.125}$$

6.14 Y In and out states

The Y fields have no interactions and are thus free fields in the model Lagrangian under consideration and in the Two-Tier quantum field theories that we construct later. Therefore "in" Y quanta are the same as "out" Y quanta.

Since the Lagrangians that we consider do not have interaction terms explicitly containing Y field factors, the S matrix is "block diagonal" in the sense that if an in-state does not contain Y quanta, (or Y coherent states) then out-states will not contain Y quanta (or coherent Y states). The proof is based on the expansion of S matrix elements using Wick's theorem in products of time ordered products of pairs of in field operators. The following equations:

$$<0|\,T(\phi_{in}(X(y_1))Y^j(y_2))|0> = 0 \tag{6.126}$$

and

$$<0|T(\phi_{in}(X(y_1))e^{-\mathbf{q}\cdot\mathbf{Y}^-(y)/M_c^2})|0> = <0|T(\phi_{in}(X(y_1))e^{+\mathbf{q}\cdot\mathbf{Y}^+(y)/M_c^2})|0> = 0 \tag{6.127}$$

prove S matrix elements with no incoming Y quanta or coherent states will have zero matrix elements to produce outgoing Y quanta or coherent states. In addition any non-zero S matrix element with n incoming Y quanta must have n outgoing Y quanta. For example an incoming state with 5 Y quanta and 2 ϕ particles can only become an outgoing state with 5 Y quanta and two or more ϕ particles. Therefore we have proved the general result:

Theorem 6.I: ***Any non-zero S matrix element has the same number of incoming Y quanta and outgoing Y quanta.***

This theorem is true in any Two-Tier quantum field theory. In order to have a tractable theory we will require all in-states and out-states not to contain Y quanta or coherent states. All normal in-state and out-state particles will contain factors of $:e^{\pm p\cdot Y/M_c^2}:$ *in the fourier expansions of their corresponding fields.*

6.15 Unitarity

For many years it has been evident that modified field theories[17,38] might offer some hope of avoiding the divergences of conventional quantum field theory. Usually

[38] S. Blaha, Phys.Rev. **D10**, 4268 (1974); S. Blaha, Phys.Rev. **D11**, 2921 (1975); S. Blaha, Nuovo Cim. **A49**, :113 (1979); S. Blaha, "Generalization of Weyl's Unified Theory to Encompass a Non-Abelian Internal Symmetry Group" SLAC-PUB-1799, Aug 1976; S. Blaha, "Quantum Gravity and Quark Confinement" Lett. Nuovo Cim. **18**, 60 (1977); S. Blaha, "The Local Definition of Asymptotic Particle States" Nuovo Cim. **A49**, 35 (1979) and references therein.

these theories suffer from unitarity problems: negative norms and negative probabilities. In the absence of a physically acceptable interpretation of negative probabilities, these theories have been thought to be unsatisfactory.

The Two-Tier type of quantum field theory *superficially* also appears to have a unitarity problem due to the non-Hermitian nature of Two-Tier Hamiltonians. The lack of hermiticity is due entirely to the appearance of iY^μ in the X^μ field coordinates. *In fact Two-Tier quantum field theories satisfy unitarity for physical states. Physical states are defined to consist of any number of normal Two-Tier particles and no Y quanta.*

Two-Tier interaction Hamiltonians are not Hermitian. For example,

$$H_{Fint} = \int d^3y' \mathscr{H}_{Fint}(\phi_{in}(\, y' + iY(y')/M_c^2)) \qquad (6.128)$$

and

$$H_{Fint}^{\dagger} = \int d^3y' \mathscr{H}_{Fint}(\phi_{in}(\, y' - iY(y')/M_c^2)) \neq H_{Fint} \qquad (6.129)$$

The relation between H_{Fint} and its Hermitian conjugate is

$$H_{Fint} = VH_{Fint}^{\dagger}V \qquad (6.130)$$

where $V^2 = I$ is the metric operator defined in eqs. 6.15 – 6.18. Thus the S matrix is not unitary; the S matrix is *pseudo-unitary*:

$$S^{-1} = VS^{\dagger}V \qquad (6.131)$$

and so

$$VS^{\dagger}VS = I \qquad (6.132)$$

We will now show that the S matrix is *unitary between physical states.* To prove this point, consider eq. 6.132 between physical states |i> and <f| – each consisting of a number of ϕ particles and no Y quanta.

$$\delta_{fi} = \langle f|I|i\rangle = \langle f|VS^{\dagger}VS|i\rangle$$

$$= \sum_{n,\,m,\,p} \langle f|V|p\rangle\langle p|S^{\dagger}|n\rangle\langle n|V|m\rangle\langle m|S|i\rangle$$

$$= \sum_{n,\,m,\,p} \langle f|S^{\dagger}|m\rangle\langle m|S|i\rangle \qquad (6.133)$$

since V has the eigenvalue 1 between states consisting of no Y quanta. If there are no incoming Y quanta, then there are no outgoing Y quanta.

The block diagonality of S (and the diagonality of V) limits the intermediate states |n> and |m> to states containing ϕ particles and no Y quanta – although normalization factors R(**p**, z) will appear (described later) due to the presence of $:e^{\pm p\cdot Y/M_c^2}:$ factors within quantum field fourier expansions that embody Y coherent state effects. Thus

$$S_{phys}^{\dagger}\, S_{phys} = I \tag{6.134}$$

and

$$S_{phys}^{\dagger} = S_{phys}^{-1} \tag{6.135}$$

proving unitarity for physical states – states consisting of ϕ particles and no Y quanta that are properly normalized. A detailed example is presented later.

6.15.1 Finite Renormalization of External Legs

In the previous section we showed the theory satisfies unitarity for states that are properly normalized. However the use of the non-unitary operator W(y) (eq. 6.6) to transform $\Phi_{in}(y)$ fields into $\phi_{in}(X(y))$ fields in the LSZ procedure in eq. 6.69, and related equations, does not preserve the norm of input and output ϕ particle legs. Thus a finite renormalization is needed for each external particle leg in order to have a unitary S-matrix.

We define this renormalization of external legs within the framework of a perturbation theory example later.

6.16 Perturbation Expansion

Perturbation theory in Two-Tier quantum field theory is very similar to conventional perturbation theory. The difference is in the form of the propagators, which have a high energy damping factor R(**p**, z) that eliminates infinities that normally appear at high energy in conventional quantum field theories.

In order to develop a feeling for Two-Tier perturbation theory we will calculate a few low order diagrams in the perturbation theory of the model scalar ϕ^4 theory that we have been using as an example in this chapter.

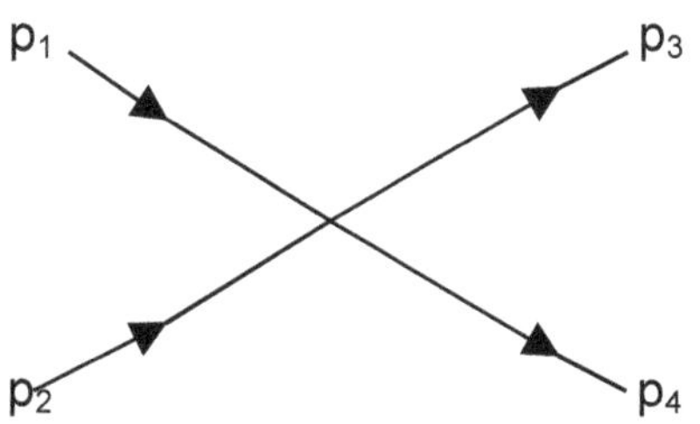

Figure 6.1. Lowest order quartic interaction diagram.

Fig. 6.1 contains the lowest order diagram for the scattering of two ϕ particles into a two ϕ particle out-state. The S matrix element for this diagram is

$$S_1 = i^4(\tfrac{1}{4!}\, i\varkappa_0) \prod_{j=1}^{4} \int d^4y_j\, d^4y\, f_{Zp_1}(y_1) f_{Zp_2}(y_2) f_{Zp_3}{}^*(y_3) f_{Zp_4}{}^*(y_4)(\Box_{y_1} + m^2)\cdot$$
$$\cdot(\Box_{y_2}+m^2)(\Box_{y_3}+m^2)(\Box_{y_4}+m^2)\langle 0|T(\phi_{in}(X(y_1))\ldots\phi_{in}(X(y_4)):(\phi_{in}(X(y))^4:)|0\rangle \qquad (6.136)$$

with $f_{Zp}(y)$ specified by

$$f_{Zp}(y) = [(2\pi)^3 2p^0 Z_p]^{-\frac{1}{2}}\, e^{-ip\cdot y} \qquad (6.137)$$

where Z_p is a normalization factor that will be specified later.

Expanding the time ordered product, and realizing there are 4! ways of combining the four field factors in the interaction Hamiltonian, we find:

$$S_1 = i^4(i\varkappa_0) \prod_{j=1}^{4} \int d^4y_j\, d^4y\, f_{Zp_1}(y_1) f_{Zp_2}(y_2) f_{Zp_3}{}^*(y_3) f_{Zp_4}{}^*(y_4)(\Box_{y_1} + m^2)\cdot$$
$$\cdot(\Box_{y_2}+m^2)(\Box_{y_3}+m^2)(\Box_{y_4}+m^2) i\Delta_F{}^{TT}(y_1-y) i\Delta_F{}^{TT}(y_2-y)\, i\Delta_F{}^{TT}(y_3-y) i\Delta_F{}^{TT}(y_4-y) \qquad (6.138)$$

where

$$i\Delta_F{}^{TT}(y_1 - y_2) = \langle 0|T(\phi(X(y_1)),\phi(X(y_2)))|0\rangle \qquad (6.139)$$

$$= i \int \frac{d^4p\, e^{-ip\cdot(y_1 - y_2)}\, R(\mathbf{p}, y_1 - y_2)}{(2\pi)^4\, (p^2 - m^2 + i\varepsilon)}$$

with

$$R(\mathbf{p}, y_1 - y_2) = \exp[-p^i p^j \Delta_{Tij}(y_1 - y_2)/M_c{}^4] \qquad (6.140)$$

(summations are over space indices only in the Y Coulomb gauge) and

$$\Delta_{Tij}(z) = \int d^3k\, e^{-ik\cdot z}(\delta_{ij} - k_i k_j/\mathbf{k}^2)/[(2\pi)^3 2\omega_k] \qquad (6.141)$$

From chapter 5 we have:

$$R(\mathbf{p}, y_1 - y_2) = \exp\{-p^2[A(v) + B(v)\cos^2\theta] / [4\pi^2 M_c{}^4 z^2]\} \qquad (6.142)$$

with

$$z^\mu = y_1{}^\mu - y_2{}^\mu \qquad (6.143)$$
$$z = |\mathbf{z}| = |\mathbf{y_1} - \mathbf{y_2}|$$

$$p = |\mathbf{p}|$$
$$v = |z^0|/ z$$

$$A(v) = (1 - v^2)^{-1} + .5v \ln[(v - 1)/(v + 1)]$$
$$B(v) = v^2(1 - v^2)^{-1} - 1.5v \ln[(v - 1)/(v + 1)]$$
$$\mathbf{p{\cdot}z} = pz \cos\theta$$

and with $|\mathbf{p}|$ denoting the length of the spatial vector **p**, while $|z^0|$ is the absolute value of z^0.

We note

$$R(\mathbf{p}, y) = R(\mathbf{p}, -y) \tag{6.144}$$

for later use.

Letting $y_i = w_i + y$ yields

$$S_1 = i^4(i\varkappa_0)(2\pi)^4\, \delta^4(p_3 + p_4 - p_1 - p_2)N^+(p_4)N^+(p_3)N(p_2)N(p_1) \tag{6.145}$$

where

$$N(p) = iZ_p^{-½}\int d^4w\ f_p(w)(\Box + m^2)\Delta_F^{TT}(w) \tag{6.146}$$

$$N^+(p) = iZ_p^{-½}\int d^4w\ f_p{*}(w)(\Box + m^2)\Delta_F^{TT}(w) \tag{6.147}$$

are "normalizations" of the "external legs" – the in and out states due to the Y field cloud around each particle with $Z^{-½}$ a renormalization factor to be determined later. In the limit of low momentum ($p \ll M_C$):

$$N(p) = N^+(p) \rightarrow -iZ_p^{-½}[(2\pi)^3\, 2p^0\,]^{-½} \tag{6.148}$$

which the reader will note is the standard normalization factor for external scalar field legs in conventional quantum field theory modulo the $Z_p^{-½}$ factor. The factor $Z_p^{-½}$ performs the finite renormalization of external legs discussed in the preceding unitarity discussion.

6.16.1 Higher Order Diagram Containing a Loop

We will now consider the simplest one loop scattering diagrams in the scalar ϕ^4 theory.

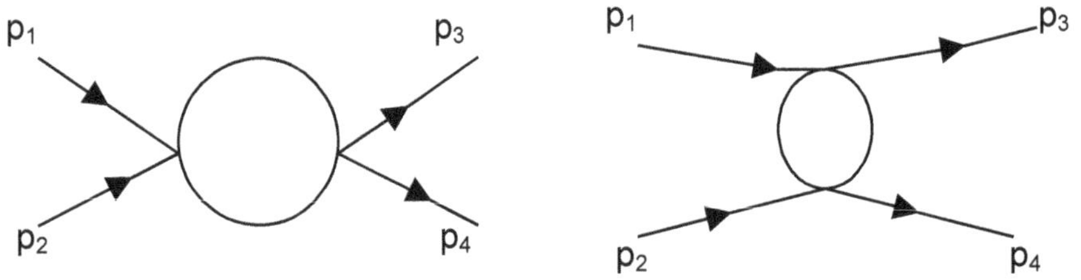

Figure 6.2. Low order loop scattering diagrams.

The S matrix element for these diagrams (and some other disconnected diagrams) is contained in

$$S_2 = i^4(\tfrac{1}{4!}\, i\varkappa_0)^2\prod_{j=1}^{4}\int d^4y_j d^4y'_1 d^4y'_2 f_{Zp_1}(y_1)f_{Zp_2}(y_2)f_{Zp_3}{}^*(y_3)f_{Zp_4}{}^*(y_4)(\Box_{y_1} + m^2)\cdot$$
$$\cdot(\Box_{y_2}+m^2)(\Box_{y_3}+m^2)(\Box_{y_4}+m^2)<0|T(\phi_{in}(X(y_1))\ldots\phi_{in}(X(y_4)) \cdot$$
$$\cdot:(\phi_{in}(X(y'_1))^4::(\phi_{in}(X(y'_2))^4:)|0>/2! \qquad (6.149)$$

together with contributions from some other disconnected diagrams.

Expanding the time ordered product and keeping only the terms corresponding to Fig. 6.2 gives:

$$S_2 = i^4(i\varkappa_0)^2/2 \prod_{j=1} \int d^4y_j\, d^4y'_1\, d^4y'_2\, f_{Zp_1}(y_1)f_{Zp_2}(y_2)f_{Zp_3}{}^*(y_3)f_{Zp_4}{}^*(y_4)\cdot$$
$$\cdot(\Box_{y_1} + m^2)\,(\Box_{y_2}+m^2)(\Box_{y_3}+m^2)(\Box_{y_4}+m^2)\cdot$$
$$\cdot\{i\Delta_F{}^{TT}(y_1-y'_1)i\Delta_F{}^{TT}(y_2-y'_1)i\Delta_F{}^{TT}(y_3-y'_2)i\Delta_F{}^{TT}(y_4-y'_2) +$$
$$+\, i\Delta_F{}^{TT}(y_1-y'_1)i\Delta_F{}^{TT}(y_2-y'_2)i\Delta_F{}^{TT}(y_3-y'_1)i\Delta_F{}^{TT}(y_4-y'_2)\}\cdot$$
$$\cdot i\Delta_F{}^{TT}(y'_1-y'_2)i\Delta_F{}^{TT}(y'_1-y'_2) \qquad (6.150)$$

Following a similar procedure to the previous calculation yields

$$S_2 = i^4[(i\varkappa_0)^2/2](2\pi)^4\delta^4(p_3 + p_4 - p_1 - p_2)N^+(p_4)N^+(p_3)N(p_2)N(p_1)\cdot$$
$$\cdot \int d^4z\, [e^{-i(p_1 + p_2)\cdot z} + e^{-i(p_1 - p_3)\cdot z}]\, [i\Delta_F{}^{TT}(z)]^2 \qquad (6.151)$$

revealing a similar normalization of the external legs to that found earlier, and also a momentum conserving delta function. The loop integrals have the form:

$$I(q) = \int d^4z\, e^{-iq\cdot z}\, [i\Delta_F^{TT}(z)]^2 \tag{6.152}$$

The behavior of the Two-Tier Feynman propagator $\Delta_F^{TT}(z)$ was studied at long and short distances previously. The large distance behavior of the Two-Tier Feynman propagator $\Delta_F^{TT}(z)$ approaches the behavior of the conventional Feynman propagator since

$$R(\mathbf{p}, z) \rightarrow 1 \tag{6.153}$$

as $z^2 = z^\mu z_\mu$ becomes much larger than M_c^{-2} ($z^2 \gg M_c^{-2}$). Thus I(q) approaches the standard one loop expression of conventional field theory at large distance (or small momentum). Again we see that Two-Tier *quantum field theory realizes a form of Correspondence Principle by approaching conventional quantum field theory at large distances.*

At short distances the Gaussian factor R(**p**, z) dominates. The Two-Tier Feynman propagator $\Delta_F^{TT}(z)$ is radically different from the conventional Feynman propagator at very short distances (very high momentum). The singular behavior of the conventional Feynman propagator is replaced with a well-behaved, high-energy (short distance) behavior. Near the light cone $M_c^{-2} \gg z^2 \rightarrow 0$ (or $p^2 \gg M_c^2$) we can approximate:

$$i\Delta_F^{TT}(z) \approx \int d^3p\, [N(p)]^2\, R(\mathbf{p}, z) \tag{6.154}$$

since $e^{-ip\cdot z}$ is approximately unity for small z. We assume the mass of the ϕ particle is negligible on this scale. Upon performing the integrations we find:

$$i\Delta_F^{TT}(z) \rightarrow \pi\, M_c^4 |z^2|/8$$

as $z^2 = z^\mu z_\mu \rightarrow 0$ from the space-like or time-like side of the light cone where | | represents the absolute value.

Therefore I(q) is finite and well-behaved. At high energy ($q^2 \gg M_c^2$)

$$I(q) \sim q^{-8}$$

since the fourier transform of $\Delta_F^{TT}(z)$ (momentum space) is

$$\Delta_F^{TT}(p) = \int d^4z\, e^{-ip\cdot z}\, \Delta_F^{TT}(z) \sim p^{-6}$$

for large p ($p^2 \gg M_c^2$). (Compare the preceding high energy behavior of I(q) with the conventional logarithmically divergent one loop result $I(q) \sim \ln(q^2/\Lambda^2)$ with Λ a cutoff.)

Thus Two-Tier quantum provides the benefits of a higher derivative theory without its drawbacks.

6.17 Finite Renormalization of External Particle Legs & Unitarity Example

The renormalization factor $Z_p^{-½}$ appearing in earlier equations is due to the use of the non-unitary operator W(y) to transform $\Phi_{in}(y)$ fields into $\phi_{in}(X(y))$ fields in the LSZ procedure. W(y) does not preserve the norm of input and output ϕ particle legs. $Z_p^{½}$ performs a finite renormalization for each external particle leg to compensate for the effects of W(y).

The required renormalization is nicely illustrated by considering the unitarity sum in the imaginary part of the preceding example.

The transition matrix T_{fi} is defined in terms of the S matrix by

$$S_{fi} = \delta_{fi} - i\,(2\pi)^4\,\delta^4(P_f - P_i)T^{(+)}{}_{fi}$$

The unitarity condition is

$$T^{(+)}{}_{fi} - T^{(-)}{}_{fi} = -i\sum_n (2\pi)^4\,\delta^4(P_n - P_i)\,T^{(-)}{}_{fn}\,T^{(+)}{}_{ni} \qquad (6.155)$$

Therefore we see that the first term on the right side of eq. 6.155 gives a transition matrix term:

$$T^{(+)}{}_{2a} = -\,i[\varkappa_0{}^2/2]N^+(p_4)N^+(p_3)N(p_2)N(p_1)\int d^4z\; e^{-iP\cdot z}[i\Delta_F{}^{TT}(z)]^2 \qquad (6.156)$$

where $P = p_1 + p_2$. Substituting for $i\Delta_F{}^{TT}$ we find that the imaginary part of $T^{(+)}{}_{2a}$ is given by

$$T^{(+)}{}_{2a} - T^{(-)}{}_{2a} = -i[\varkappa_0{}^2/2]N^+(p_4)N^+(p_3)N(p_2)N(p_1)\int d^4z\; e^{-iP\cdot z}\,\cdot$$
$$\cdot\,[i\int d^4p\,\theta(p_0)\,\delta(p^2 - m^2)e^{-ip\cdot z}\,R(\mathbf{p}, z)/(2\pi)^3]^2$$

Note $R(\mathbf{p}, z)$ is real.

If we express the $R(\mathbf{p}, z)$ factors in terms of their fourier transforms:

$$R(\mathbf{p}, z) = \int d^4q\; e^{-iq\cdot z}\,R(\mathbf{p}, q)$$

Then we find

$$T^{(+)}{}_{2a} - T^{(-)}{}_{2a} = -i[\varkappa_0{}^2/2]N^+(p_4)N^+(p_3)N(p_2)N(p_1)\int d^4z\; e^{-iP\cdot z}\,\cdot$$
$$\cdot[i\int d^4k_1\,d^4q_1\theta(k_1{}^0)\,\delta(k_1{}^2 - m^2)e^{-ik_1\cdot z}e^{-iq_1\cdot z}\,R(\mathbf{k_1}, q_1)/(2\pi)^3]\cdot$$
$$\cdot[i\int d^4k_2\,d^4q_2\theta(k_2{}^0)\,\delta(k_2{}^2 - m^2)e^{-ik_2\cdot z}e^{-iq_2\cdot z}\,R(\mathbf{k_2}, q_2)/(2\pi)^3]$$

Performing the integral over z gives

$$T^{(+)}{}_{2a} - T^{(-)}{}_{2a} = +i[\varkappa_0{}^2/2]N^+(p_4)N^+(p_3)N(p_2)N(p_1)(2\pi)^4\cdot$$
$$\cdot\int d^4k_1d^4q_1d^4k_2d^4q_2\theta(k_1{}^0)\,\delta(k_1{}^2 - m^2)\,\theta(k_2{}^0)\,\delta(k_2{}^2 - m^2)\cdot$$
$$\cdot R(\mathbf{k_1}, q_1)R(\mathbf{k_2}, q_2)\,\delta^4(P + k_1 + q_1 + k_2 + q_2)/(2\pi)^6$$

Introducing delta functions enables us to re-express this equation as

$$T^{(+)}{}_{2a} - T^{(-)}{}_{2a} = +i[\varkappa_0{}^2/2]N^+(p_4)N^+(p_3)N(p_2)N(p_1)\int d^4r_1d^4r_2(2\pi)^4\,\delta^4(P - r_1 - r_2)\cdot$$
$$\cdot\int d^4k_1d^4q_1\,\delta^4(r_1 + k_1 + q_1)\theta(k_1{}^0)\,\delta(k_1{}^2 - m^2)\,R(\mathbf{k_1}, q_1)\cdot$$
$$\cdot\int d^4k_2d^4q_2\theta(k_2{}^0)\,\delta(k_2{}^2 - m^2)\,\delta^4(r_2 + k_2 + q_2)R(\mathbf{k_2}, q_2)/(2\pi)^6$$

which becomes

$$T^{(+)}{}_{2a} - T^{(-)}{}_{2a} = +i[\varkappa_0{}^2/2]N^+(p_4)N^+(p_3)N(p_2)N(p_1)\int d^4r_1d^4r_2(2\pi)^4\,\delta^4(P - r_1 - r_2)\cdot$$
$$\cdot\int d^4k_1\theta(k_1{}^0)\,\delta(k_1{}^2 - m^2)\,R(\mathbf{k_1}, -k_1 - r_1)\cdot$$
$$\cdot\int d^4k_2\theta(k_2{}^0)\,\delta(k_2{}^2 - m^2)R(\mathbf{k_2}, -k_2 - r_2)/(2\pi)^6$$

$R(\mathbf{k_2}, -k_2 - r_2)$ can be expressed in terms of its fourier transform $R(\mathbf{k_2}, z)$.
We can now rewrite the above expression in terms of intermediate states:

$$T^{(+)}{}_{2a} - T^{(-)}{}_{2a} = -i\int d^4r_1d^4r_2\,(2\pi)^4\delta^4(P - r_1 - r_2)\cdot$$
$$\cdot\, i\varkappa_0N^+(p_4)N^+(p_3)\int d^4k_1\theta(k_1{}^0)\delta(k_1{}^2 - m^2)[R(\mathbf{k_1}, -k_1 - r_1)/(2\pi)^3]\cdot$$
$$\cdot\int d^4k_2\theta(k_2{}^0)\delta(k_2{}^2 - m^2)[R(\mathbf{k_2}, -k_2 - r_2)/(2\pi)^3]i\,\varkappa_0N(p_2)N(p_1)/2$$

which has the form:

$$T^{(+)}{}_{2a} - T^{(-)}{}_{2a} = -i\int d^4r_1d^4r_2(2\pi)^4\delta^4(P - r_1 - r_2)\Big[\int d^4k_1\theta(k_1{}^0)\,\delta(k_1{}^2 - m^2)\cdot$$
$$\cdot R(\mathbf{k_1}, -k_1 - r_1)/(2\pi)^3\Big]\Big[\int d^4k_2\theta(k_2{}^0)\,\delta(k_2{}^2 - m^2)R(\mathbf{k_2}, -k_2 - r_2)/$$
$$/(2\pi)^3\Big]T^{(-)}{}_{fn}T^{(+)}{}_{ni}/2! \qquad (6.157)$$

where

$$T^{(-)}{}_{fn} = \varkappa_0N^+(p_4)N^+(p_3)N(r_2)N(r_1)$$

$$T^{(+)}{}_{ni} = \varkappa_0N^+(r_2)N^+(r_1)N(p_2)N(p_1)$$

if

$$N^{+}(p) = N(p) = 1 \qquad (6.158)$$

Eq. 6.74 implies the (finite) external leg renormalization must be

$$Z_p = -\left[\int d^4w\, f_p(w)(\Box + m^2)\Delta_F{}^{TT}(w)\right]^2 \qquad (6.159)$$

Thus all external legs must be "lopped off."

The result is a theory that satisfies the unitarity condition as shown in the above detailed discussion.

If we define

$$N(r) = \int d^4k\, \theta(k^0)\delta(k^2 - m^2)R(\mathbf{k}, -k - r) \qquad (6.160)$$

$$= (2\pi)^{-4}\int d^4k d^4z \theta(k^0)\delta(k^2 - m^2)e^{-i(k + r)\cdot z}R(\mathbf{k}, z) \qquad (6.161)$$

then the Two-Tier completeness expression becomes:

$$S_{fi} = \sum_n (2\pi)^{-3n}(n!)^{-1}\int\left(\prod_{j=1}^{n} d^4r_j N(r_j)\right)S_{fn}S_{ni}{}^{\dagger}\,\delta^4\left(P_n - \sum_{k=1}^{n} r_k\right) \qquad (6.162)$$

This expression reflects the fact that ϕ particles are surrounded by a "cloud" of Y quanta. Thus we have achieved unitarity! For small momenta $r_j \ll M_c$, we find $N(r_j) \simeq \theta(r_j^0)\delta(r_j^2 - m^2)$ (using $R(k, q) \simeq 1$). $\theta(r_j^0)\delta(r_j^2 - m^2)$ is the form seen in conventional quantum field theory. For large momenta $N(r_j)$ is very different.

6.18 General Form of Propagators

In this chapter we have considered a scalar Two-Tier quantum field theory. We have seen that the Two-Tier Feynman propagator is well behaved near the light cone resulting in a finite ϕ^4 theory. This finite ϕ^4 theory approximates the results of conventional ϕ^4 theory at low energy thus implementing a correspondence principle: *At low energy, results in Two-Tier quantum field theory approach the corresponding results of conventional quantum field theory.*

The observations on Two-Tier field theory made in this chapter generally apply to Two-Tier versions of Quantum Electrodynamics, ElectroWeak Theory and the Standard Model as well as Two-Tier Quantum Gravity:

1. At low energy ($p^2 \ll M_c^2$ or large distances $z^2 \gg M_c^{-2}$) the Two-Tier quantum field theory is the same as the corresponding conventional quantum field theory to good approximation. (Correspondence Principle)

2. At high energy ($p^2 \gg M_c^2$ or short distances: $z^2 \ll M_c^{-2}$) Two-Tier quantum field theories (of physical interest) are well-behaved and finite.

3. Two-Tier quantum field theories (of physical interest) satisfy unitarity and Lorentz invariance (and in the case of quantum gravity their dynamical equations satisfy the requirements of general relativity).

The generality of these results is based on:

1. The expansion of the S matrix in time ordered products of field operators.

2. Wick's Theorem

3. The general form of all particle propagators in Two-Tier quantum field theories. All particle Feynman propagators have the form:

$$iG_F{}^{TT}{}_{...}(y_1 - y_2) = \langle 0|T(\chi_{...}(X(y_1)), \chi_{...}(X(y_2)))|0\rangle \qquad (6.163)$$

$$= \int d^4p\, iG_{F...}(p)e^{-ip\cdot(y_1 - y_2)}\, R(\mathbf{p}, y_1 - y_2) \qquad (6.164)$$

where $iG_{F...}(p)$ is the conventional momentum space $\chi_{...}$ particle propagator, and where … represents the relevant tensor and matrix indices. $R(\mathbf{p}, y_1 - y_2)$ introduces a common damping factor in each particle propagator that eliminates divergences.

6.18.1 Scalar Particle Propagator

The Two-Tier propagator for the case of a free scalar particle is:

$$i\Delta_F{}^{TT}(y_1 - y_2) = \langle 0|T(\phi(X(y_1)), \phi(X(y_2)))|0\rangle \qquad (6.165)$$

$$= i \int \frac{d^4p\; e^{-ip\cdot(y_1 - y_2)}\, R(\mathbf{p}, y_1 - y_2)}{(2\pi)^4\,(p^2 - m^2 + i\varepsilon)}$$

Since the mass m is not relevant at high energy we set m = 0. This enables us to obtain a more tractable expression for the propagator. After some manipulation the massless scalar propagator can be represented as:

$$i\Delta_F{}^{TT}(z) = -\beta[16\pi^3(AB)^{1/2}]^{-1} \int_{-\infty}^{\infty} dy_1 \int_{-\infty}^{\infty} dy_2 \cdot$$

$$\cdot\{\theta(z_0)\exp[-\beta((y_1 - z_0)^2B + (y_2 + z)^2A)/(4AB)] +$$

$$+ \theta(-z_0)\exp[-\beta((y_1 + z_0)^2B + (y_2 - z)^2A)/(4AB)]\}/(y_1^2 - y_2^2) \tag{6.166}$$

with $\beta = 4\pi^2 M_c^{\,4}\mathbf{z}^2$. Using

$$(y_1^2 - y_2^2)^{-1} = -0.5 \int_0^\infty dq_1 \int_{-\infty}^\infty dq_2\ \theta(q_1^2 - q_2^2)\exp[iq_1y_1 - iq_2y_2] \tag{6.167}$$

we obtain the representation

$$i\Delta_F^{TT}(z^\mu) = (8\pi^2)^{-1} \int_0^\infty dq_1 \int_{-\infty}^\infty dq_2\ \theta(q_1^2 - q_2^2)\ \cdot$$

$$\cdot \exp\{iq_1|z_0| + iq_2 z\ - [A'q_1^2 + B'q_2^2]/[\beta'(\mathbf{z}^2 - z_0^2)]\} \tag{6.168}$$

where $|z_0|$ is the absolute value of z_0, $z^2 - z_0^2 = -z^\mu z_\mu$ and

$$A = A'/(1 - v^2) \tag{6.169}$$
$$B = B'/(1 - v^2) \tag{6.170}$$
$$\beta = 4\pi^2 M_c^{\,4}\mathbf{z}^2 = \beta' z^2 \tag{6.171}$$

with $\mathbf{z} = |\vec{\mathbf{z}}|$ – the magnitude of the spatial vector $\vec{\mathbf{z}}$, and A and B given above.

The representation of $i\Delta_F^{TT}$ in eq. 6.168 is particularly useful in determining its low energy ($\ll M_c$), and its high energy ($\gg M_c$) behavior. The low energy behavior is governed by the linear terms in the exponential in eq. 6.168 since $\beta'(\mathbf{z}^2 - z_0^2)$ is very large in this limit:

$$i\Delta_F^{TT}(z^\mu)_{low} \simeq (8\pi^2)^{-1} \int_0^\infty dq_1 \int_{-\infty}^\infty dq_2\ \theta(q_1^2 - q_2^2)\exp\{\ iq_1|z_0| + iq_2 z\} \tag{6.172}$$

$$= [4\pi^2(\mathbf{z}^2 - z_0^2)]^{-1}$$
$$= i\Delta_F(z^\mu)$$

equaling the exact massless, free, spin 0 Feynman propagator of conventional quantum field theory.

In the high energy limit when $\beta'(\mathbf{z}^2 - z_0^2)$ is small since $\mathbf{z}^2 \approx z_0^2$ (i.e. near the light cone), the quadratic terms in the exponential dominate, and $A' \simeq B'$. We then find

$$i\Delta_F^{TT}(z^\mu)_{high} \simeq (8\pi^2)^{-1} \int_0^\infty dq_1 \int_{-\infty}^\infty dq_2 \theta(q_1^2 - q_2^2)\exp\{A'(q_1^2 + q_2^2)/[\beta'(\mathbf{z}^2 - z_0^2)]\} \tag{6.173}$$

$$= \pi M_c^{\,4}|(\mathbf{z}^2 - z_0^2)|/8 \tag{6.174}$$

As pointed out earlier, eq. 6.174 corresponds to k^{-6} behavior in momentum space:

$$i\Delta_F{}^{TT}(k)_{high} \backsim k^{-6} \tag{6.175}$$

6.18.2 Spin ½ Particle Propagator

For the case of a free, spin ½ particle the propagator is:

$$iS_F{}^{TT}(y_1 - y_2) = \langle 0|T(\bar{\psi}(X(y_1))\psi(X(y_2)))|0\rangle \tag{6.176}$$
$$= i\int \frac{d^4p\, e^{-ip\cdot(y_1 - y_2)}\,(\not p + m)\, R(\mathbf{p}, y_1 - y_2)}{(2\pi)^4\,(p^2 - m^2 + i\varepsilon)}$$

Again setting m = 0 we find a convenient representation in the form:

$$S_F{}^{TT}(z^\mu) = i(8\pi^2)^{-1}\int_0^\infty dq_1\int_{-\infty}^\infty dq_2\; \theta(q_1{}^2 - q_2{}^2)(\epsilon(z_0)q_1\gamma_0 - q_2\vec{\mathbf{z}}\cdot\vec{\gamma}/z)\cdot$$
$$\cdot\exp\{iq_1|z_0| + iq_2z - [A'q_1{}^2 + B'q_2{}^2]/[\beta'(\mathbf{z}^2 - z_0{}^2)]\} \tag{6.177}$$

using the same symbols and notation as above, and with $\epsilon(z_0) = +1$ if $z_0 \geq 0$ and -1 otherwise.

The representation of $S_F{}^{TT}$ in eq. 6.177 is useful in determining its low energy ($\ll M_c$), and high energy ($\gg M_c$) behavior. The low energy behavior is governed by the linear terms in the exponential in eq. 6.177 since $\beta'(\mathbf{z}^2 - z_0{}^2)$ is large in this limit:

$$S_F{}^{TT}(z^\mu)_{low} \simeq (8\pi^2)^{-1}\int_0^\infty dq_1\int_{-\infty}^\infty dq_2\; \theta(q_1{}^2 - q_2{}^2)(\epsilon(z_0)q_1\gamma_0 - q_2\vec{\mathbf{z}}\cdot\vec{\gamma}/z)\cdot$$
$$\cdot\exp\{iq_1|z_0| + iq_2z\} \tag{6.178}$$

$$= \not z[2\pi^2(\mathbf{z}^2 - z_0{}^2)^2]^{-1} \tag{6.179}$$
$$= S_F(z^\mu)$$

equaling the exact massless, spin ½ Feynman propagator of conventional quantum field theory. If we had not set m = 0 initially, we would have obtained the usual massive, spin ½ Feynman propagator.

In the high energy limit when $\beta'(\mathbf{z}^2 - z_0{}^2)$ is small since $\mathbf{z}^2 \approx z_0{}^2$ (i.e. near the light cone), the quadratic terms in the exponential dominate, and $A' \simeq B'$. We then find

$$S_F^{TT}(z^\mu)_{high} \simeq (8\pi^2)^{-1} \int_0^\infty dq_1 \int_{-\infty}^\infty dq_2 \theta(q_1^2 - q_2^2)(\in(z_0)q_1\gamma_0 - q_2\vec{\mathbf{z}} \cdot \vec{\gamma}/z) \cdot$$

$$\cdot \exp\{A'(q_1^2 + q_2^2)/[\beta'(\mathbf{z}^2 - z_0^2)]\} \qquad (6.180)$$

$$= i(8\pi^2)^{-1}\{\mathbf{z}^{-1}(\mathbf{z}^2 - z_0^2)^{3/2}2^{3/2}\pi^{7/2}M_c^6 z_0\gamma_0 - $$
$$- 4i(\mathbf{z}^2 - z_0^2)^2\pi^5 M_c^8 \vec{\mathbf{z}} \cdot \vec{\gamma})\} \qquad (6.181)$$

The leading momentum dependence of the fourier transform of $S_F^{TT}(z^\mu)_{high}$ is

$$S_F^{TT}(p)_{high} \backsim M_c^6 p^{-7}\gamma_0 \qquad (6.182)$$

6.18.3 Massless Spin 1 Particle Propagator

The Two-Tier Feynman propagator for the case of a free, massless, spin 1, gauge field particle (coupled to a conserved current) such as a photon is:

$$iD_F^{TT}(z)_{\mu\nu} = -i \int \frac{d^4p\, e^{-ip\cdot z}\, g_{\mu\nu}\, R(\mathbf{p}, y_1 - y_2)}{(2\pi)^4\,(p^2 + i\varepsilon)} \qquad (6.183)$$

The form of eq. 6.183 is the same as the scalar particle propagator multiplied by $-g_{\mu\nu}$. As a result we have the representation:

$$iD_F^{TT}(z)_{\mu\nu} = -(8\pi^2)^{-1} \int_0^\infty dq_1 \int_{-\infty}^\infty dq_2\, \theta(q_1^2 - q_2^2)\, g_{\mu\nu} \cdot$$

$$\cdot \exp\{iq_1|z_0| + iq_2 z - [A'q_1^2 + B'q_2^2]/[\beta'(\mathbf{z}^2 - z_0^2)]\} \quad (6.184)$$

As before in the scalar particle case, the low energy behavior is governed by the linear terms in the exponential in eq. 6.96 since $\beta'(\mathbf{z}^2 - z_0^2)$ is very large in this limit:

$$iD_F^{TT}(z)_{\mu\nu low} \simeq -g_{\mu\nu}(8\pi^2)^{-1} \int_0^\infty dq_1 \int_{-\infty}^\infty dq_2\, \theta(q_1^2 - q_2^2)\exp\{iq_1|z_0| + iq_2 z\}$$

$$= -g_{\mu\nu}[4\pi^2(\mathbf{z}^2 - z_0^2)]^{-1} \qquad (6.185)$$
$$= -ig_{\mu\nu}\Delta_F(z)$$

equaling the exact free, massless, spin 1 Feynman gauge field propagator of conventional quantum field theory.

In the high energy limit when $\beta'(\mathbf{z}^2 - z_0^2)$ is small since $\mathbf{z}^2 \approx z_0^2$ (i.e. near the light cone), the quadratic terms in the exponential dominate, and $A' \simeq B'$. We then find

$$iD_F^{TT}(z)_{\mu\nu hi} \simeq -(8\pi^2)^{-1}\int_0^\infty dq_1 \int_{-\infty}^\infty dq_2 \theta(q_1^2 - q_2^2) g_{\mu\nu} \exp\{A'(q_1^2 + q_2^2)/[\beta'(\mathbf{z}^2 - z_0^2)]\} \tag{6.186}$$

$$= -g_{\mu\nu}\pi M_c^4 |(\mathbf{z}^2 - z_0^2)|/8 \tag{6.187}$$

Eq. 6.187 corresponds to k^{-6} behavior in momentum space:

$$iD_F^{TT}(k)_{\mu\nu high} \sim g_{\mu\nu} M_c^4 k^{-6} \tag{6.188}$$

6.18.4 Spin 2 Particle Propagator

The Two-Tier propagator for the case of a free, massless, spin 2 particle such as a graviton is:

$$i\Delta_{F2}^{TT}(z)_{\mu\nu\rho\sigma} = i\int \frac{d^4p\, e^{-ip\cdot z}\, b_{\mu\nu\rho\sigma}(p) R(\mathbf{p}, y_1 - y_2)}{(2\pi)^4 (p^2 + i\varepsilon)} \tag{6.189}$$

in an appropriate gauge where $b_{\mu\nu\rho\sigma}(p)$ is a tensor that is independent of the coordinates. We can express eq. 6.189 in the form:

$$i\Delta_{F2}^{TT}(z)_{\mu\nu\rho\sigma} = (8\pi^2)^{-1}\int_0^\infty dq_1 \int_{-\infty}^\infty dq_2\, \theta(q_1^2 - q_2^2)\, \tilde{b}(z_0, z, q_1, q_2)_{\mu\nu\rho\sigma} \cdot$$
$$\cdot \exp\{ iq_1|z_0| + iq_2 z - [A'q_1^2 + B'q_2^2]/[\beta'(\mathbf{z}^2 - z_0^2)]\} \tag{6.190}$$

where $\tilde{b}(z_0, z, q_1, q_2)_{\mu\nu\rho\sigma}$ is a tensor generated from the $b_{\mu\nu\rho\sigma}(p)$ tensor.

As before in the scalar particle case, the low energy behavior is governed by the linear terms in the exponential since $\beta'(\mathbf{z}^2 - z_0^2)$ is very large in this limit and we find that the covariant piece[39] behaves like:

$$i\Delta_{F2}{}^{TT}(z)_{\mu\nu\rho\sigma\text{lowCov}} \simeq \tilde{\tilde{b}}_{\mu\nu\rho\sigma}(8\pi^2)^{-1}\int_0^\infty dq_1 \int_{-\infty}^\infty dq_2\ \theta(q_1{}^2 - q_2{}^2)\exp\{\ iq_1|z_0| + iq_2 z\} \tag{6.191}$$

$$= \tilde{\tilde{b}}_{\mu\nu\rho\sigma}[4\pi^2(\mathbf{z}^2 - z_0{}^2)]^{-1} \tag{6.192}$$

$$= i\Delta_F(z^\mu)\ \tilde{\tilde{b}}_{\mu\nu\rho\sigma}$$

where

$$\tilde{\tilde{b}}_{\mu\nu\rho\sigma} = \tfrac{1}{2}\,[\eta_{\mu\rho}\eta_{\nu\sigma} + \eta_{\mu\sigma}\eta_{\nu\rho} - \eta_{\mu\nu}\eta_{\rho\sigma}] \tag{6.193}$$

so that the expression in eq. 6.192 equals the corresponding covariant piece of the exact free, massless, spin 2 Feynman propagator of conventional quantum field theory.

In the high energy limit when $\beta'(\mathbf{z}^2 - z_0{}^2)$ is small since $\mathbf{z}^2 \approx z_0{}^2$ (i.e. near the light cone), the quadratic terms in the exponential dominate, and $A' \simeq B'$. We then find

$$i\Delta_{F2}{}^{TT}(z)_{\mu\nu\rho\sigma\text{high}} \simeq (8\pi^2)^{-1}\int_0^\infty dq_1 \int_{-\infty}^\infty dq_2\theta(q_1{}^2 - q_2{}^2)\ \tilde{b}(z_0, z, q_1, q_2)_{\mu\nu\rho\sigma}\cdot$$

$$\cdot\ \exp\{A'(q_1{}^2 + q_2{}^2)/[\ \beta'(\mathbf{z}^2 - z_0{}^2)]\} \tag{6.194}$$

and the covariant piece behaves like

$$i\Delta_{F2}{}^{TT}(z)_{\mu\nu\rho\sigma\text{highCov}} \simeq \tilde{\tilde{b}}_{\mu\nu\rho\sigma}\pi M_c{}^4|(\mathbf{z}^2 - z_0{}^2)|/8 \tag{6.195}$$

The coordinate space behavior corresponds to k^{-6} behavior in momentum space:

$$i\Delta_{F2}{}^{TT}(k)_{\mu\nu\rho\sigma\text{highCov}} \backsim \tilde{\tilde{b}}_{\mu\nu\rho\sigma}\, k^{-6} \tag{6.196}$$

The high-energy behavior of the spin 2 propagator in momentum space results in a Two-Tier theory of quantum gravity that has no high-energy divergences and is thus *finite*.

---

39 S. Weinberg, Phys. Rev. **135**, B1049 (1964); Phys. Rev. **138**, B988 (1965).

REFERENCES

Akhiezer, N. I., Frink, A. H. (tr), 1962, *The Calculus of Variations* (Blaisdell Publishing, New York, 1962).

Bjorken, J. D., Drell, S. D., 1964, *Relativistic Quantum Mechanics* (McGraw-Hill, New York, 1965).

Bjorken, J. D., Drell, S. D., 1965, *Relativistic Quantum Fields* (McGraw-Hill, New York, 1965).

Blaha, S., 1995, *C++ for Professional Programming* (International Thomson Publishing, Boston, 1995).

_______, 1998, *Cosmos and Consciousness* (Pingree-Hill Publishing, Auburn, NH, 1998 and 2002).

_______, 2002, *A Finite Unified Quantum Field Theory of the Elementary Particle Standard Model and Quantum Gravity Based on New Quantum Dimensions™ & a New Paradigm in the Calculus of Variations* (Pingree-Hill Publishing, Auburn, NH, 2002).

_______, 2004, *Quantum Big Bang Cosmology: Complex Space-time General Relativity, Quantum Coordinates™ Dodecahedral Universe, Inflation, and New Spin 0, ½, 1 & 2 Tachyons & Imagyons* (Pingree-Hill Publishing, Auburn, NH, 2004).

______, *2005a, Quantum Theory of the Third Kind: A New Type of Divergence-free Quantum Field Theory Supporting a Unified Standard Model of Elementary Particles and Quantum Gravity based on a New Method in the Calculus of Variations* (Pingree-Hill Publishing, Auburn, NH, 2005).

______, 2005b, *The Metatheory of Physics Theories, and the Theory of Everything as a Quantum Computer Language* (Pingree-Hill Publishing, Auburn, NH, 2005).

______, 2005c, *The Equivalence of Elementary Particle Theories and Computer Languages: Quantum Computers, Turing Machines, Standard Model, Superstring Theory, and a Proof that Gödel's Theorem Implies Nature Must Be Quantum* (Pingree-Hill Publishing, Auburn, NH, 2005).

______, 2006a, *The Foundation of the Forces of Nature* (Pingree-Hill Publishing, Auburn, NH, 2006).

______, 2006b, *A Derivation of ElectroWeak Theory based on an Extension of Special Relativity; Black Hole Tachyons; & Tachyons of Any Spin.* (Pingree-Hill Publishing, Auburn, NH, 2006).

_______, 2007a, *Physics Beyond the Light Barrier: The Source of Parity Violation, Tachyons, and A Derivation of Standard Model Features* (Pingree-Hill Publishing, Auburn, NH, 2007).

______, 2007b, *The Origin of the Standard Model: The Genesis of Four Quark and Lepton Species, Parity Violation, the ElectroWeak Sector, Color SU(3), Three Visible Generations of Fermions, and One Generation of Dark Matter with Dark Energy* (Pingree-Hill Publishing, Auburn, NH, 2007).

______, *2008a, A Direct Derivation of the Form of the Standard Model From GL(16) (Pingree-Hill Publishing, Auburn, NH, 2008).*

______, 2008b, *A Complete Derivation of the Form of the Standard Model With a New Method to Generate Particle Masses Second Edition* (Pingree-Hill Publishing, Auburn, NH, 2008)

______, 2009, *The Algebra of Thought & Reality: The Mathematical Basis for Plato's Theory of Ideas, and Reality Extended to Include A Priori Observers and Space-Time Second Edition* (Pingree-Hill Publishing, Auburn, NH, 2009).

______, 2010a, *Operator Metaphysics: A New Metaphysics Based on a New Operator Logic and a New Quantum Operator Logic that Lead to a Mathematical Basis for Plato's Theory of Ideas and Reality* (Pingree-Hill Publishing, Auburn, NH, 2010).
______, 2010b, *The Standard Model's Form Derived from Operator Logic, Superluminal Transformations and GL(16)* (Pingree-Hill Publishing, Auburn, NH, 2010).

______, 2010c, *SuperCivilizations: Civilizations as Superorganisms* (McMann-Fisher Publishing, Auburn, NH, 2010).

______, 2011a, *21st Century Natural Philosophy Of Ultimate Physical Reality* (McMann-Fisher Publishing, Auburn, NH, 2011).

______, 2011b, *All the Universe! Faster Than Light Tachyon Quark Starships & Particle Accelerators with the LHC as a Prototype Starship Drive Scientific Edition* (Pingree-Hill
Publishing, Auburn, NH, 2011).

______, 2011c, *From Asynchronous Logic to The Standard Model to Superflight to the Stars* (Blaha Research, Auburn, NH, 2011).

______, 2012a, *From Asynchronous Logic to The Standard Model to Superflight to the Stars volume 2: Superluminal CP and CPT, U(4) Complex General Relativity and The Standard Model, Complex Vierbein General Relativity, Kinetic Theory, Thermodynamics* (Blaha Research, Auburn, NH, 2012).

______, 2012b, *Standard Model Symmetries, And Four And Sixteen Dimension Complex Relativity; The Origin Of Higgs Mass Terms* (Blaha Reasearch, Auburn, NH, 2012).

______, 2013a, *Multi-Stage Space Guns, Micro-Pulse Nuclear Rockets, and Faster-Than-Light Quark-Gluon Ion Drive Starships* (Blaha Research, Auburn, NH, 2013).

______, 2013b, *The Bridge to Dark Matter; A New Sibling Universe; Dark Energy; Inflatons; Quantum Big Bang; Superluminal Physics; An Extended Standard Model Based on Geometry* (Blaha Reasearch, Auburn, NH, 2013).

______, 2014a, *Universes and Megaverses: From a New Standard Model to a Physical Megaverse; The Big Bang; Our Sibling Universe's Wormhole; Origin of the Cosmological Constant, Spatial Asymmetry of the Universe, and its Web of Galaxies; A Baryonic Field between Universes and Particles; Megaverse Extended Wheeler-DeWitt Equation* (Blaha Reasearch, Auburn, NH, 2014).

______, 2014b, *All the Megaverse! Starships Exploring the Endless Universes of the Cosmos Using the Baryonic Force* (Blaha Research, Auburn, NH, 2014).

______, 2014c, *All the Megaverse! II Between Megaverse Universes: Quantum Entanglement Explained by the Megaverse Coherent Baryonic Radiation Devices – PHASERs Neutron Star Megaverse Slingshot Dynamics Spiritual and UFO Events, and the Megaverse Microscopic Entry into the Megaverse* (Blaha Research, Auburn, NH, 2014).

______, 2015a, *PHYSICS IS LOGIC PAINTED ON THE VOID: Origin of Bare Masses and The Standard Model in Logic, U(4) Origin of the Generations, Normal and Dark Baryonic Forces, Dark Matter, Dark Energy, The Big Bang, Complex General Relativity, A Megaverse of Universe Particles* (Blaha Research, Auburn, NH, 2015).

______, 2015b, *PHYSICS IS LOGIC Part II: The Theory of Everything, The Megaverse Theory of Everything, U(4)⊗U(4) Grand Unified Theory (GUT), Inertial Mass = Gravitational Mass, Unified Extended Standard Model and a New Complex General Relativity with Higgs Particles, Generation Group Higgs Particles* (Blaha Research, Auburn, NH, 2015).

______, 2015c, *The Origin of Higgs ("God") Particles and the Higgs Mechanism: Physics is Logic III, Beyond Higgs – A Revamped Theory With a Local Arrow of Time, The Theory of Everything Enhanced, Why Inertial Frames are Special, Universes of the Mind* (Blaha Research, Auburn, NH, 2015).

______, 2015d, *The Origin of the Eight Coupling Constants of The Theory of Everything: U(8) Grand Unified Theory of Everything (GUTE), S^8 Coupling Constant Symmetry, Space-Time Dependent Coupling Constants, Big Bang Vacuum Coupling Constants, Physics is Logic IV* (Blaha Research, Auburn, NH, 2015).

______, 2016a, *New Types of Dark Matter, Big Bang Equipartition, and A New U(4) Symmetry in the Theory of Everything: Equipartition Principle for Fermions, Matter is 83.33% Dark, Penetrating the Veil of the Big Bang, Explicit QFT Quark Confinement and Charmonium, Physics is Logic V* (Blaha Research, Auburn, NH, 2016).

______, 2016b, *The Periodic Table of the 192 Quarks and Leptons in The Theory of Everything: The U(4) Layer Group, Physics is Logic VI* (Blaha Research, Auburn, NH, 2016).

______, 2016c, *New Boson Quantum Field Theory, Dark Matter Dynamics, Dark Matter Fermion Layer Mixing, Genesis of Higgs Particles, New Layer Higgs Masses, Higgs Coupling Constants, Non-Abelian Higgs Gauge Fields, Physics is Logic VII* (Blaha Research, Auburn, NH, 2016).

______, 2016d, *Unification of the Strong Interactions and Gravitation: Quark Confinement Linked to Modified Short-Distance Gravity; Physics is Logic VIII* (Blaha Research, Auburn, NH, 2016).

______, 2016e, *MoND: Unification of the Strong Interactions and Gravitation II, Quark Confinement Linked to Large-Scale Gravity, Physics is Logic IX* (Blaha Research, Auburn, NH, 2016).

______, 2016f, *CQ Mechanics: A Unification of Quantum & Classical Mechanics, Quantum/Semi-Classical Entanglement, Quantum/Classical Path Integrals, Quantum/Classical Chaos* (Blaha Research, Auburn, NH, 2016).

______, 2016g, *GEMS Unified Gravity, ElectroMagnetic and Strong Interactions: Manifest Quark Confinement, A Solution for the Proton Spin Puzzle, Modified Gravity on the Galactic Scale* (Pingree Hill Publishing, Auburn, NH, 2016).

______, 2016h, *Unification of the Seven Boson Interactions based on the Riemann-Christoffel Curvature Tensor* (Pingree Hill Publishing, Auburn, NH, 2016).

______, 2017a, *Unification of the Eleven Boson Interactions based on 'Rotations of Interactions'* (Pingree Hill Publishing, Auburn, NH, 2017).

______, 2017b, *The Origin of Fermions and Bosons, and Their Unification* (Pingree Hill Publishing, Auburn, NH, 2017).

______, 2017c, *Megaverse: The Universe of Universes* (Pingree Hill Publishing, Auburn, NH, 2017).

______, 2017d, *SuperSymmetry and the Unified SuperStandard Model* (Pingree Hill Publishing, Auburn, NH, 2017).

______, 2017e, *From Qubits to the Unified SuperStandard Model with Embedded SuperStrings: A Derivation* (Pingree Hill Publishing, Auburn, NH, 2017).

______, 2017f, *The Unified SuperStandard Model in Our Universe and the Megaverse: Quarks, …* , (Pingree Hill Publishing, Auburn, NH, 2017).

______, 2018a, *The Unified SuperStandard Model and the Megaverse SECOND EDITION A Deeper Theory based on a New Particle Functional Space that Explicates Quantum Entanglement Spookiness (Volume 1)* (Pingree Hill Publishing, Auburn, NH, 2018).

______, 2018b, *Cosmos Creation: The Unified SuperStandard Model, Volume 2, SECOND EDITION* (Pingree Hill Publishing, Auburn, NH, 2018).

______, 2018c, *God Theory (*Pingree Hill Publishing, Auburn, NH, 2018).

______, 2018d, *Immortal Eye: God Theory: Second Edition* (Pingree Hill Publishing, Auburn, NH, 2018).
______, 2018e, *Unification of God Theory and Unified SuperStandard Model THIRD EDITION* (Pingree Hill Publishing, Auburn, NH, 2018).

______, 2019a, *Calculation of: QED α = 1/137, and Other Coupling Constants of the Unified SuperStandard Theory* (Pingree Hill Publishing, Auburn, NH, 2019).

______, 2019b, *Coupling Constants of the Unified SuperStandard Theory SECOND EDITION* (Pingree Hill Publishing, Auburn, NH, 2019).

______, 2019c, *New Hybrid Quantum Big_Bang–Megaverse_Driven Universe with a Finite Big Bang and an Increasing Hubble Constant* (Pingree Hill Publishing, Auburn, NH, 2019).

______, 2019d, *The Universe, The Electron and The Vacuum* (Pingree Hill Publishing, Auburn, NH, 2019).

______, 2019e, *Quantum Big Bang – Quantum Vacuum Universes (Particles)* (Pingree Hill Publishing, Auburn, NH, 2019).

______, 2019f, *The Exact QED Calculation of the Fine Structure Constant Implies ALL 4D Universes have the Same Physics/Life Prospects* (Pingree Hill Publishing, Auburn, NH, 2019).

______, 2019g, *Unified SuperStandard Theory and the SuperUniverse Model: The Foundation of Science* (Pingree Hill Publishing, Auburn, NH, 2019).

______, 2020a, *Quaternion Unified SuperStandard Theory (The QUeST) and Megaverse Octonion SuperStandard Theory (MOST)* (Pingree Hill Publishing, Auburn, NH, 2020).

______, 2020b, *United Universes Quaternion Universe - Octonion Megaverse* (Pingree Hill Publishing, Auburn, NH, 2020).

______, 2020c, *Unified SuperStandard Theories for Quaternion Universes & The Octonion Megaverse* (Pingree Hill Publishing, Auburn, NH, 2020).

______, 2020d, *The Essence of Eternity: Quaternion & Octonion SuperStandard Theories* (Pingree Hill Publishing, Auburn, NH, 2020).

______, 2020e, *The Essence of Eternity II* (Pingree Hill Publishing, Auburn, NH, 2020).

______, 2020f, *A Very Conscious Universe* (Pingree Hill Publishing, Auburn, NH, 2020).

______, 2020g, *Hypercomplex Universe* (Pingree Hill Publishing, Auburn, NH, 2020).

______, 2020h, *Beneath the Quaternion Universe* (Pingree Hill Publishing, Auburn, NH, 2020).

______, 2020i, *Why is the Universe Real? From Quaternion & Octonion to Real Coordinates* (Pingree Hill Publishing, Auburn, NH, 2020).

______, 2020j, *The Origin of Universes: of Quaternion Unified SuperStandard Theory (QUeST); and of the Octonion Megaverse (UTMOST)* (Pingree Hill Publishing, Auburn, NH, 2020).

______, 2020k, *The Seven Spaces of Creation: Octonion Cosmology* (Pingree Hill Publishing, Auburn, NH, 2020).

______, 2020l, *From Octonion Cosmology to the Unified SuperStandard Theory of Particles* (Pingree Hill Publishing, Auburn, NH, 2020).

______, 2021a, *Pioneering the Cosmos* (Pingree Hill Publishing, Auburn, NH, 2021).

______, 2021b, *Pioneering the Cosmos II* (Pingree Hill Publishing, Auburn, NH, 2021).

______, 2021c, *Beyond Octonion Cosmology* (Pingree Hill Publishing, Auburn, NH, 2021).

______, 2021d, *Universes are Particles* (Pingree Hill Publishing, Auburn, NH, 2021).

______, 2021e, *Octonion-like dna-based life, Universe expansion is decay, Emerging New Physics* (Pingree Hill Publishing, Auburn, NH, 2021).

______, 2021f, *The Science of Creation New Quantum Field Theory of Spaces* (Pingree Hill Publishing, Auburn, NH, 2021).

______, 2021g, *Quantum Space Theory With Application to Octonion Cosmology & Possibly To Fermionic Condensed Matter* (Pingree Hill Publishing, Auburn, NH, 2021).

______, 2021h, *21st Century Natural Philosophy of Octonion Cosmology , and Predestination, Fate, and Free Will* (Pingree Hill Publishing, Auburn, NH, 2021).

______, 2021i, *Beyond Octonion Cosmology II : Origin of the Quantum; A New Generalized Field Theory (GiFT); A Proof of the Spectrum of Universes; Atoms in Higher Universes* (Pingree Hill Publishing, Auburn, NH, 2021).

______, 2021j, *Integration of General Relativity and Quantum Theory: Octonion Cosmology, GiFT, Creation/Annihilation Spaces CASe, Reduction of Spaces to a Few Fermions and Symmetries in Fundamental Frames* (Pingree Hill Publishing, Auburn, NH, 2021).

______, 2022a, *New View of Octonion Cosmology Based on the Unification of General Relativity and Quantum Theory* (Pingree Hill Publishing, Auburn, NH, 2022).

______, 2022b, *The Dust Beneath Hypercomplex Cosmology* (Pingree Hill Publishing, Auburn, NH, 2022).

______, 2022c, *Passing Through Nature to Eternity: ProtoCosmos, HyperCosmos, Unified SuperStandard Theory* (Pingree Hill Publishing, Auburn, NH, 2022).

______, 2022d, *HyperCosmos Fractionation and Fundamental Reference Frame Based Unification: Particle Inner Space Basis of Parton and Dual Resonance Models* (Pingree Hill Publishing, Auburn, NH, 2022).

______, 2022e, *A New UniDimension ProtoCosmos and SuperString F-Theory Relation to the HyperCosmos* (Pingree Hill Publishing, Auburn, NH, 2022).

______, 2022f, *The Cosmic Panorama: ProtoCosmos, HyperCosmos, Unified SuperStandard Theory (UST) Derivation* (Pingree Hill Publishing, Auburn, NH, 2022).

______, 2022g, *Ultimate Origin: ProtoCosmos and HyperCosmos* (Pingree Hill Publishing, Auburn, NH, 2022).

______, 2023a, *UltraUnification and the Generation of the Cosmos* (Pingree Hill Publishing, Auburn, NH, 2023).

______, 2023b, *God and and Cosmos Theory* (Pingree Hill Publishing, Auburn, NH, 2023).

______, 2023c, *A New Completely Geometric SU(8) Cosmos Theory; New PseudoFermion Fields; Fibonacci-like Dimension Arrays; Ramsey Number Approximation* (Pingree Hill Publishing, Auburn, NH, 2023).

______, 2023d, *Newton's Apple is Now the Fermion* (Pingree Hill Publishing, Auburn, NH, 2023).

REFERENCES

______, 2023e,*Cosmos Theory: The Sub-Particle Gambol Model* (Pingree Hill Publishing, Auburn, NH, 2023).

_____, 2024a, *Cosmos-Universe-Particle-Gambol Theory* (Pingree Hill Publishing, Auburn, NH, 2024).

_____, 2024b, *Fractal Cosmos Theory* (Pingree Hill Publishing, Auburn, NH, 2024).

_____, 2024c, *Fractal Cosmic Curve: Tensor-Based CosmosTheory* (Pingree Hill Publishing, Auburn, NH, 2024).
_____, 2024d, *The Eternal Form of Cosmos Theory* (Pingree Hill Publishing, Auburn, NH, 2024).

_____, 2024e, *The Eternal Form of Cosmos Theory Third Edition* (Pingree Hill Publishing, Auburn, NH, 2024).

_____, 2024f, *Fundamental Constants of Cosmos Theory and The Standard Model* (Pingree Hill Publishing, Auburn, NH, 2024).

_____, 2024g, *Quark, Lepton, W and Z Masses of Cosmos Theory and The Standard Model* (Pingree Hill Publishing, Auburn, NH, 2024).

_____, 2024h, *Geometric Cosmos Geometric Universe* (Pingree Hill Publishing, Auburn, NH, 2024).

_____, 2024i, *Particles and Universes of Cosmos Theory* (Pingree Hill Publishing, Auburn, NH, 2024).

_____, 2024j, *Unification of the Subluminal and the Superluminal in Cosmos Theory* (Pingree Hill Publishing, Auburn, NH, 2024).

Eddington, A. S., 1952, *The Mathematical Theory of Relativity* (Cambridge University Press, Cambridge, U.K., 1952).

Fant, Karl M., 2005, *Logically Determined Design: Clockless System Design With NULL Convention Logic* (John Wiley and Sons, Hoboken, NJ, 2005).

Feinberg, G. and Shapiro, R., 1980, *Life Beyond Earth: The Intelligent Earthlings Guide to Life in the Universe* (William Morrow and Company, New York, 1980).

Gelfand, I. M., Fomin, S. V., Silverman, R. A. (tr), 2000, *Calculus of Variations* (Dover Publications, Mineola, NY, 2000).

Giaquinta, M., Modica, G., Souchek, J., 1998, *Cartesian Coordinates in the Calculus of Variations* Volumes I and II (Springer-Verlag, New York, 1998).

Giaquinta, M., Hildebrandt, S., 1996, *Calculus of Variations* Volumes I and II (Springer-Verlag, New York, 1996).

Gradshteyn, I. S. and Ryzhik, I. M., 1965, *Table of Integrals, Series, and Products* (Academic Press, New York, 1965).

Heitler, W., 1954, *The Quantum Theory of Radiation* (Claendon Press, Oxford, UK, 1954).

Huang, Kerson, 1992, *Quarks, Leptons & Gauge Fields 2nd Edition* (World Scientific Publishing Company, Singapore, 1992).

Jost, J., Li-Jost, X., 1998, *Calculus of Variations* (Cambridge University Press, New York, 1998).

Kaku, Michio, 1993, *Quantum Field Theory*, (Oxford University Press, New York, 1993).

Kirk, G. S. and Raven, J. E., 1962, *The Presocratic Philosophers* (Cambridge University Press, New York, 1962).

Landau, L. D. and Lifshitz, E. M., 1987, *Fluid Mechanics 2nd Edition*, (Pergamon Press, Elmsford, NY, 1987).

REFERENCES

Rescher, N., 1967, *The Philosophy of Leibniz* (Prentice-Hall, Englewood Cliffs, NJ, 1967).

Riesz, Frigyes and Sz.-Nagy, Béla, 1990, *Functional Analysis* (Dover Publications, New York, 1990).

Sakurai, J. J., 1964, *Invariance Principles and Elementary Particles* (Princeton University Press, Princeton, NJ, 1964).

Weinberg, S., 1972, *Gravitation and Cosmology* (John Wiley and Sons, New York, 1972).

Weinberg, S., 1995, *The Quantum Theory of Fields Volume I* (Cambridge University Press, New York, 1995).

REFERENCES

INDEX

About the Author

Stephen Blaha is a well-known Physicist and Man of Letters with interests in Science, Society and civilization, the Arts, and Technology. He had an Alfred P. Sloan Foundation scholarship in college. He received his Ph.D. in Physics from Rockefeller University. He has served on the faculties of several major universities. He was also a Member of the Technical Staff at Bell Laboratories, a manager at the Boston Globe Newspaper, a Director at Wang Laboratories, and President of Blaha Software Inc. and of Janus Associates Inc. (NH).

Among other achievements he was a co-discoverer of the "r potential" for heavy quark binding developing the first (and still the only demonstrable) non-Aeolian gauge theory with an "r" potential; first suggested the existence of topological structures in superfluid He-3; first proposed Yang-Mills theories would appear in condensed matter phenomena with non-scalar order parameters; first developed a grammar-based formalism for quantum computers and applied it to elementary particle theories; first developed a new form of quantum field theory without divergences (thus solving a major 60 year old problem that enabled a unified theory of the Standard Model and Quantum Gravity without divergences to be developed); first developed a formulation of complex General Relativity based on analytic continuation from real space-time; first developed a generalized non-homogeneous Robertson-Walker metric that enabled a quantum theory of the Big Bang to be developed without singularities at t = 0; first generalized Cauchy's theorem and Gauss' theorem to complex, curved multi-dimensional spaces; received Honorable Mention in the Gravity Research Foundation Essay Competition in 1978; first developed a physically acceptable theory of faster-than-light particles; first derived a composition of extremums method in the Calculus of Variations; first quantitatively suggested that inflationary periods in the history of the universe were not needed; first proved Gödel's Theorem implies Nature must be quantum; provided a new alternative to the Higgs Mechanism, and Higgs particles, to generate masses; first showed how to resolve logical paradoxes including Gödel's Undecidability Theorem by developing Operator Logic and Quantum Operator Logic; first developed a quantitative harmonic oscillator-like model of the life cycle, and interactions, of civilizations; first showed how equations describing superorganisms also apply to civilizations. A recent book shows his theory applies successfully to the past 14 years of history and to *new* archaeological data on Andean and Mayan civilizations as well as Early Anatolian and Egyptian civilizations.

He first developed an axiomatic derivation of the form of The Standard Model from geometry – space-time properties – The Unified SuperStandard Model. It unifies all the known forces of Nature. It also has a Dark Matter sector that includes a Dark ElectroWeak sector with Dark doublets and Dark gauge interactions. It uses quantum coordinates to remove infinities that crop up in most interacting quantum field theories and additionally to remove the infinities that appear in the Big Bang and generate inflationary growth of the universe. It shows gravity has a MOND-like form without sacrificing Newton's Laws. It relates the interactions of the MOND-like sector of gravity with the r-potential of Quark Confinement. The axioms of the theory lead to the question of their origin. We suggest in the preceding edition of this book it can be attributed to an entity with God-like properties. We explore these properties in "God Theory" and show they predict that the Cosmos exists forever although individual universes (or incarnations of our universe) "come and go." Several other important results emerge from God Theory such a functionally triune God. The Unified SuperStandard Theory has many other important parts described in the Current Edition of *The Unified SuperStandard Theory* and expanded in subsequent volumes.

Blaha has had a major impact on a succession of elementary particle theories: his Ph.D. thesis (1970), and papers, showed that quantum field theory calculations to all orders in ladder approximations could not give scaling deep inelastic electron-nucleon scattering. He later showed the eigenvalue equation for the fine structure constant α in Johnson-Baker-Willey QED had a zero at $\alpha = 1$ not 1/137 by solving the Schwinger-Dyson equations to all orders in an approximation that agreed with exact results to 4th order in α thus ending interest in this theory. In 1979 at Prof. Ken Johnson's (MIT) suggestion he calculated the proton-neutron mass difference in the MIT bag model and found the result had the wrong sign reducing interest in the bag model. These results all appear in Physical Review papers. In the 2000's he repeatedly pointed out the shortcomings of SuperString theory and showed that The Standard Model's form could be derived from space-time geometry by an extension of Lorentz transformations to faster than light transformations. This deeper space-time basis greatly increases the possibility that it is part of THE fundamental theory. Recently, Blaha showed that the Weak interactions differed significantly from the Strong, electromagnetic and gravitation interactions in important respects while these interactions had similar features, and suggested that ElectroWeak theory, which is essentially a glued union of the Weak interactions and Electromagnetism, possibly modulo unknown Higgs particle features, be replaced by a unified theory of the other interactions combined with a stand-alone Weak interaction theory. Blaha also showed that, if Charmonium calculations are taken seriously, the Strong interaction coupling constant is only a factor of five larger than the electromagnetic coupling constant, and thus Strong interaction perturbation theory would make sense and yield physically meaningful results.

In graduate school (1965-71) he wrote substantial papers in elementary particles and group theory: The Inelastic E- P Structure Functions in a Gluon Model. Phys. Lett. B40:501-502,1972; Deep-Inelastic E-P Structure Functions In A Ladder Model With Spin 1/2 Nucleons, Phys.Rev. D3:510-523,1971; Continuum Contributions To The Pion Radius, Phys. Rev. 178:2167-2169,1969; Character Analysis of U(N) and SU(N), J. Math. Phys. 10, 2156 (1969); and The Calculation of the Irreducible Characters of the Symmetric Group in Terms of the Compound Characters, (Published as Blaha's Lemma in D. E. Knuth's book: *The Art of Computer Programming Vols. 1 – 4*).

In the early 1980's Blaha was also a pioneer in the development of UNIX for financial, scientific and Internet applications: benchmarked UNIX versions showing that block size was critical for UNIX performance, developing financial modeling software, starting database benchmarking comparison studies, developing Internet-like UNIX networking (1982) and developing a hybrid shell programming technique (1982) that was a precursor to the PERL programming language. He was also the manager of the AT&T ten-year future products development database. His work helped lead to commercial UNIX on computers such as Sun Micros, IBM AIX minis, and Apple computers.

In the 1980's he pioneered the development of PC Desktop Publishing on laser printers and was nominated for three "Awards for Technical Excellence" in 1987 by PC Magazine for PC software products that he designed and developed.

Recently he has developed a theory of Megaverses – actual universes of which our universe is one – with quantum particle-like properties based on the Wheeler-DeWitt equation of Quantum Gravity. He has developed a theory of a baryonic force, which had been conjectured many years ago, and estimated the strength of the force based on discrepancies in measurements of the gravitational constant G. This force, operative in D-dimensional space, can be used to escape from our universe in "uniships" which are the equivalent of the faster-than-light starships proposed in the author's earlier books. Thus travel to other universes, as well as to other stars is possible.

Blaha also considered the complexified Wheeler-DeWitt equation and showed that its limitation to real-valued coordinates and metrics generated a Cosmological Constant in the Einstein equations.

The author has also recently written a series of books on the serious problems of the United States and their solution as well as a book on the decline of Mankind that will follow from current social and genetic trends in Mankind.

In the past twenty years Dr. Blaha has written over 80 books on a wide range of topics. Some recent major works are: *From Asynchronous Logic to The Standard Model to Superflight to the Stars, All the Universe!, SuperCivilizations: Civilizations as Superorganisms, America's Future: an Islamic Surge, ISIS, al Qaeda, World Epidemics, Ukraine, Russia-China Pact, US Leadership Crisis, The Rises and Falls of Man – Destiny – 3000 AD: New Support for a Superorganism MACRO-THEORY of CIVILIZATIONS From CURRENT WORLD TRENDS and NEW Peruvian, Pre-Mayan, Mayan, Anatolian, and Early Egyptian Data, with a Projection to 3000 AD,* and *Mankind in Decline: Genetic Disasters, Human-Animal Hybrids, Overpopulation, Pollution, Global Warming, Food and Water Shortages, Desertification, Poverty, Rising Violence, Genocide, Epidemics, Wars, Leadership Failure.*

He has taught approximately 4,000 students in undergraduate, graduate, and postgraduate corporate education courses primarily in major universities, and large companies and government agencies.

He developed a quantum theory, The Unified SuperStandard Theory (UST), which describes elementary particles in detail without the difficulties of conventional quantum field theory. He found that the internal symmetries of this theory could be exactly derived from an octonion theory called QUeST. He further found that another octonion theory (UTMOST) describes the Megaverse. It can hold QUeST universes such as our own universe. It has an internal symmetry structure which is a superset of the QUeST internal symmetries.

Recently he developed Octonion Cosmology. He replaced it with HyperCosmos theory, which has significantly better features. He developed a fractionalization process for dimensions, particles and symmetry groups. He also described transformation that reduced particles and dimensions to a far more compact form. He also developed a precursor theory ProtoCosmos that leads to the HyperCosmos.

The author showed that space-time and Internal Symmetries can be unified in any of the ten HyperCosmos spaces in their associated HyperUnification spaces. The combined set of HyperUnification spaces enable all HyperCosmos dimensions to be obtained by a General Relativistic transformation from one primordial dimension in the 42 space-time dimension unified HyperUnification space.

At present the author devel;oped the Cosmos Theory that incorporates ProtoCosmos Theory, HyperCosmos Theory, Limos Theory, Second Kind HyperCosmos Theory and HyperUnification Spaces. He has introduced PseudoFermion wave functions and theory, He has related Cosmos Theory to Regge trajectories of spaces, parton theory, Veneziano amplitudes, Fibonacci numbers and Ramsey numbers. He has calculated an approximation to the difficult R(n,n) Ramsey numbers.

He has developed a Gambol Model that successfully accounts for e-p deep inelastic scattering, fundamental particle resonances, hadron scattering, and the inner structure of particles based on confinement through Casimir forces of ideal gambol gases. The Gambol Planckian Distribution was derived.

He has applied the Gambol Model to particles, universes, and the Cosmos of universes. He showed that the Cosmos may have a distribution of 23 universes corresponding to various Cosmos spaces.

Recently he showed that Cosmos Theory follows from the number of independent asymmetric tensors in a dimension r. He also showed the close parallel between the form of γ-matrices and Cosmos Theory dimension arrays. The closeness suggested that dimension arrays have the same importance as γ-matrices for fermions.

He demonstrated that the pressure of fermions within a space of dimension r balances the Casimir vacuum energy force for 18 dimensions. He showed that $2e\pi$ = 17.02 marks the critical point where pressure balances Casimir force, which implies r = 18 is the highest dimension Physical Cosmos space. The dimension $2e\pi$ appears to set the approximate dimension for Cosmos spaces with dimension array size $2^{r+4} \cong (17.02/8)^{r+4}. \cong (e\pi/4)^{r+4}. \cong 2.13^{r+4}$.

Now he has found the sequence of Coupling Constant values in the Standard Model and UST.

Milton Keynes UK
Ingram Content Group UK Ltd.
UKHW052111011124
450603UK00005B/45

9 798989 408481